£5.95
-T6D.
SC

ROYAL SOCIETY OF CHEMISTRY PAPERBACKS

D1280832

Food
the chemistry of its components

T. P. COULTATE MIBiol, PhD, AIFST
Polytechnic of the South Bank, London

LONDON: THE ROYAL SOCIETY OF CHEMISTRY

85 0000 2855

© The Royal Society of Chemistry, 1984

ISBN 0 85186 483 X

All Rights Reserved. No part of this book may be reproduced or transmitted in any form or by any means – graphic, electronic, including photocopying, recording, taping or information storage and retrieval systems – without written permission from The Royal Society of Chemistry.

Published by The Royal Society of Chemistry, Burlington House, London W1V 0BN, and distributed by The Royal Society of Chemistry Distribution Centre, Blackhorse Road, Letchworth, Herts SG6 1HN

Typeset by Unicus Graphics Ltd, Horsham
Printed by Whitstable Litho Ltd

TX
531
.C68
1984

Contents

Acknowledgements

I am indebted to my colleagues in the Department of Applied Biology and Food Science for their encouragement to write this book and their forbearance with the difficulties that have arisen because I have been writing it when I should have been administrating.

Similarly my wife and sons must be thanked for putting up with an even smaller share of my time than is usual for the families of polytechnic lecturers.

Thanks are especially due to my colleague and commuting companion, Dr David Rosie, who not only checked the manuscript but also together with my long suffering office companion, Mr Ken Spears, and many classes of my students, acted as a testing ground for much of the material of this book. Of course the opinions that are expressed, and any errors that remain are mine, not theirs.

I would also like to thank Mrs Valerie Everett, Mrs Patricia Smith and Miss Karen Heritage for their contributions to the typing, and Mr Melvyn Stevens for obtaining the spectra in *Fig. 5.12*.

1. Introduction

For the chemists of the 18th and 19th centuries an understanding of the chemical nature of our food was a major objective. They realised that this knowledge was essential if dietary standards, and with them health and prosperity, were to improve. Inevitably it was the food components present in large amounts, the carbohydrates, fats and proteins that were first to be described. As physiologists and physicians began to relate their findings to the chemical knowledge of foodstuffs the greatest need became one for analytical techniques, a demand no less pressing today. The food components that occur in much smaller amounts, the pigments, vitamins and flavour compounds for example, required 20th century laboratory techniques for their isolation and characterisation. Thus in spite of the more or less similar importance of the different classes of food components the extent of knowledge has not advanced evenly.

By the time of World War II it appeared that most of the questions being asked of food chemists by nutritionists, agriculturalists and others had been answered. This was certainly true as far as questions of the 'what is this substance and how much is there?' variety were concerned. However, as reflected in this book, over the past 20 years or so new questions have been asked, and so far only a few answers have been obtained. Food chemists nowadays are required to explain the *behaviour* of food components - on storage, processing, cooking, even in the mouth and during digestion. Much of the stimulus to this type of enquiry has come from the food manufacturing industry - and the legislative bodies which attempt to control the industry's activities. For example the observation that the starch in a dessert product provides a certain amount of energy has been overtaken in importance by the need to know which type of starch will give just the right degree of thickening, and what is the molecular basis for the differences between one starch and another.

The later chapters of this book show that with regard to the quantitatively less prominent components the examination of their properties in food systems is only just beginning. With the obvious exception of the vitamins this delay has been caused, at least in part, by the failure of nutritionists, physiologists and other scientists to recognise what housewives* and the food manufacturing industry have always known. That is that there is more to the business of feeding people than compiling a list of nutrients in the correct proportions. Furthermore, this is as true if one is engaged in famine relief as it is in a five star restaurant. To satisfy a nutritional need a foodstuff must be acceptable, and to be acceptable it must first look and then taste 'right'.

The search for the answers to questions of food texture, colour and flavour as well as simple composition have turned the chemical study of food into a mongrel discipline. Its present vigour, which stimulated the writing of this book, comes from the necessary integration of normally separate scientific disciplines. For example in Chapter 4 the chemistry of meat is shown to require a knowledge of cell biology allied to free radical chemistry. Similarly in Chapter 8 it would have been foolish to consider the chemistry of preservatives without some insights from food microbiology.

With such a far ranging subject no book of this size can claim to be comprehensive. It is ironic that the most abundant food component of all, water, has not been give the chapter to itself that it deserves. Other omissions, particularly of many aspects of nutrition, have been provoked by the abundance of excellent textbooks, at all academic levels, rather than any indifference on the part of the author. A number of excellent texts covering related areas of food science have been listed in Appendix 1.

No attempt has been made to make this a manual of laboratory experiments, although frequent references to laboratory methods have been made. In the area of practical food chemistry there is no substitute for 'Pearson' (see Appendix 1). In view of the increasing use of sophisticated instrumental methods which may well be beyond the scope of school or college laboratories there is much to be said for one of the earlier editions of Pearson's book (*eg* the 5th or 6th), written in the days when the burette still reigned unchallenged in the analytical laboratory.

---

*Throughout this book the term 'housewife' is to be taken to mean anyone, of either sex, married or single, responsible for the retail purchase of food and its preparation in the home.

2. Carbohydrates

Sugars and their derivatives

Sugars, such as sucrose and glucose, together with polysaccharides such as starch and cellulose, are the principal components of the class of substances we call *carbohydrates*. Although chemists never seem to have difficulty in deciding whether or not a particular substance should be classified as a carbohydrate they have been unable to provide a concise, formal definition. The empirical formulae of most of the carbohydrates we encounter in foodstuffs approximate to $(CH_2O)_n$, hence the name. More usefully it is simpler to regard them as aliphatic polyhydroxy compounds which usually carry a carbonyl group (and, of course, derivatives of such compounds).

The simplest carbohydrates are the monosaccharides. These have between three and eight carbon atoms but only those with five or six are common. The suffix '-ose' is included in the names of monosaccharides that have a carbonyl group and in the absence of any other identification the number of carbon atoms is indicated by terms such as triose, tetrose and pentose. The prefixes aldo- and keto- are used to show whether the carbonyl group is at the first or a subsequent carbon atom so that we may refer to, for example, aldohexoses or ketopentoses. To complicate matters further the two triose monosaccharides are almost never named in this way but are referred to as glyceraldehyde (2,3-dihydroxypropanal) [2.1] and dihydroxyacetone (dihydroxypropanone) [2.2].

2.1	2.2
CHO	CH_2OH
|	|
CHOH	CO
|	|
CH_2OH	CH_2OH

Of greatest concern to us will be the aldo- [2.3] and [2.5] and keto- [2.4] and [2.6] pentoses [2.3] and [2.4] and hexoses [2.5] and [2.6], shown here with the conventional numbering of the carbon atoms:

2.3	2.4	2.5	2.6
^{1}CHO	$^{1}CH_2OH$	^{1}CHO	$^{1}CH_2OH$
$^{2}CHOH$	^{2}CO	$^{2}CHOH$	^{2}CO
$^{3}CHOH$	$^{3}CHOH$	$^{3}CHOH$	$^{3}CHOH$
$^{4}CHOH$	$^{4}CHOH$	$^{4}CHOH$	$^{4}CHOH$
$^{5}CH_2OH$	$^{5}CH_2OH$	$^{5}CHOH$	$^{5}CHOH$
		$^{6}CH_2OH$	$^{6}CH_2OH$

The carbon atom of each CHOH is of course asymmetrically substituted and all carbohydrates show optical isomerism. Almost all naturally occurring monosaccharides belong to the so-called D-series. That is to say their highest numbered asymmetric carbon, the one furthest from the carbonyl group, has the same configuration as D-glyceraldehyde [2.7] rather than its isomer L-glyceraldehyde [2.8].

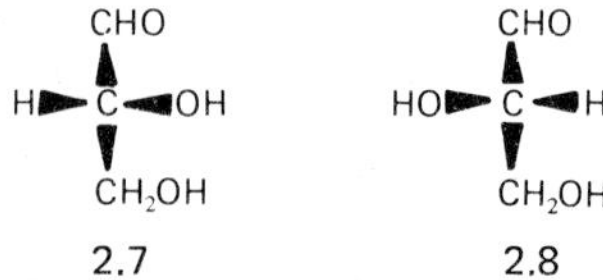

2.7 2.8

The structural relationships of the higher aldose monosaccharides are shown in *Fig. 2.1*. For simplicity in this diagram and most subsequent structural formulae the carbon atoms within the chain are indicated by the intersections of the vertical and horizontal bonds. The optical configuration corresponds to the conventional representation of the asymmetric carbon atom of glyceraldehyde shown above. The names of the numerous optical isomers of the aldoses will best be remembered by use of the awful mnemonics 'Get Raxl!' and 'All altruists gladly make gum in gallon tanks' (I would be delighted to hear of any improved versions). A corresponding table of ketose sugars may be drawn up but with the exception of D-fructose [2.9] none of the ketoses are of much significance to food chemists.

It is important to remember that the monosaccharides of the L-series are related to L-glyceraldehyde [2.8] and have the mirror image configurations to the corresponding D-series sugars. Thus L-glucose is [2.10] rather than [2.11] which is in fact L-idose. It is not surprising, in view of their asymmetry, that the monosaccharides are optically active, *ie* their solutions and crystals rotate the plane of polarized light. The symbols (+) and (−) can be used to denote rotation to the right and left respectively. The old fashioned names of dextrose and laevulose

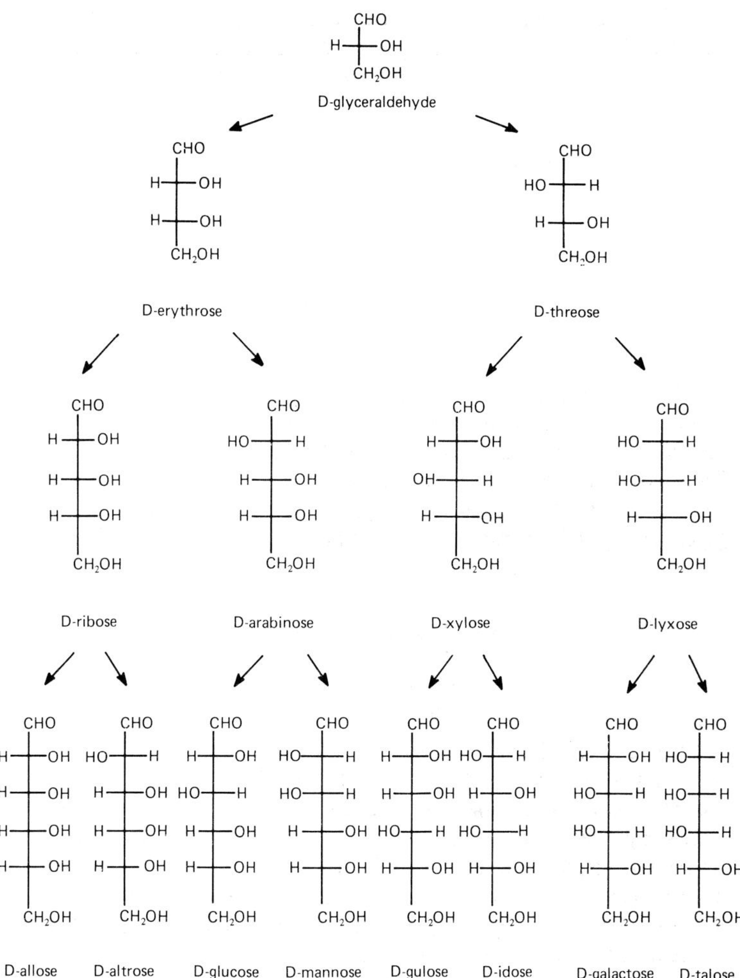

FIG. 2.1. The configurations of the D-aldoses

2.9 2.10 2.11

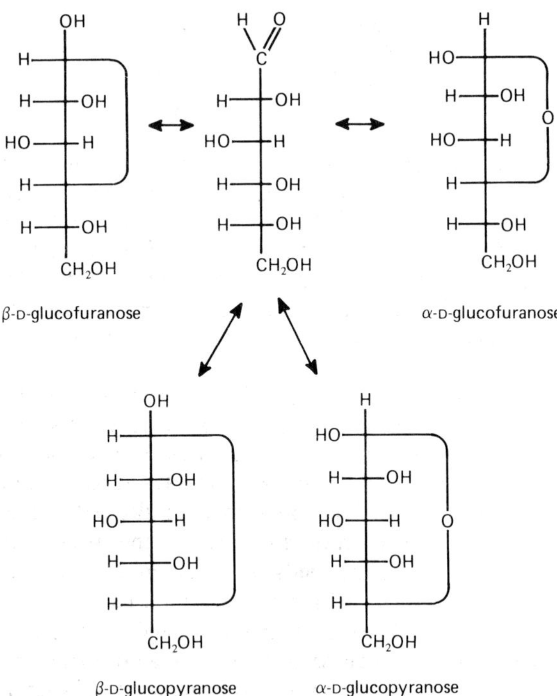

FIG. 2.2. Hemiacetal ring structures and mutarotation in D-glucose. Only the pyranose forms of D-glucose actually occur in significant amounts. The actual mechanism of mutarotation is not fully established. It occurs most rapidly at extremes of pH and apparently involves concerted acidic and basic catalysis.

---

stem from the respectively dextrotatory and laevorotatory properties of D(+)-glucose and D(−)-fructose.

The straight chain structural formulae that have been used so far in this account do not satisfactorily account for many monosaccharide properties. In particular their reactions, while showing them to possess a carbonyl group, are not entirely typical of carbonyl compounds generally. The differences are explained by the formation of a ring structure by a condensation of the carbonyl group with a hydroxyl group at the far end of the chain to form a *hemiacetal*. As *Fig. 2.2* shows, there are four possible ring structures for D-glucose, differing in the configuration of the new asymmetric centre and the size of the ring. When the ring is six membered, as in *Fig. 2.2*, it is referred to as *pyranose* and when five membered *furanose*, from the structures of pyran [2.12] and furan [2.13] respectively. The two isomeric forms of the particular monosaccharide are known as *anomers* and are designated α and β.

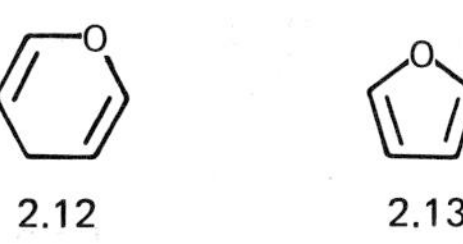

The α anomer is the one that has the hydroxyl group derived from the reducing group (the anomeric hydroxyl) on the opposite face of the ring to carbon six (see *Fig. 2.2*). The α and β anomers of a particular monosaccharide will differ in optical rotation. For example, α-D-glucopyranose, which is the form in which D-glucose crystallises from aqueous solution, has a specific rotation* of +112° whereas that of β-D-glucose, which is the form which crystallises from pyridine solution, is +19°.

When crystals of either α- or β-D-glucose are dissolved in water the specific rotation is observed to change until, regardless of the form one started with, the solution finally gives a value of +52°. This phenomenon is known as *mutarotation*. The transition of one anomer to another proceeds through the open chain or *aldehydo-* form and it is clear that it is this isomer which is involved in the sugar reactions that are typical of carbonyl compounds, even though in aqueous solutions only 0.02 per cent of D-glucose molecules are in this form.

The structural relationships between L and D isomers, α and β anomers, pyranose and furanose rings in both aldose and ketose sugars are by no means easily mastered. If available the use of a set of molecular models will help to clarify the issues but the structural formulae set out in *Fig. 2.3* illustrate the essential features of terminology in this area.

Over the years chemists have synthesised innumerable derivatives of monosaccharides but only a few occur naturally or have particular significance to food. Oxidation of the carbonyl group of aldose sugars leads to the formation of the '-onic' series of sugar acids. Thus the enzyme glucose oxidase catalyses the formation of D-gluconolactone [2.14] which hydrolyses spontaneously to D-gluconic acid [2.15]. This enzyme is highly specific for the β anomer of D-glucose and forms the basis of a popular analytical technique for glucose determination in food materials as well as a process for removing traces of glucose from the bulk liquid egg used in commercial bakeries and elsewhere (to prevent the Maillard reaction – see page 23).

---

*The specific rotation at a temperature of 20 °C using light of the D line of the sodium spectrum is given by the expression

$$[\alpha]_D^{20} = \frac{100\alpha}{l \times c}$$

where α is the rotation observed in a polarimeter tube of length l decimeters and a sugar concentration of c grams per 100 ml.

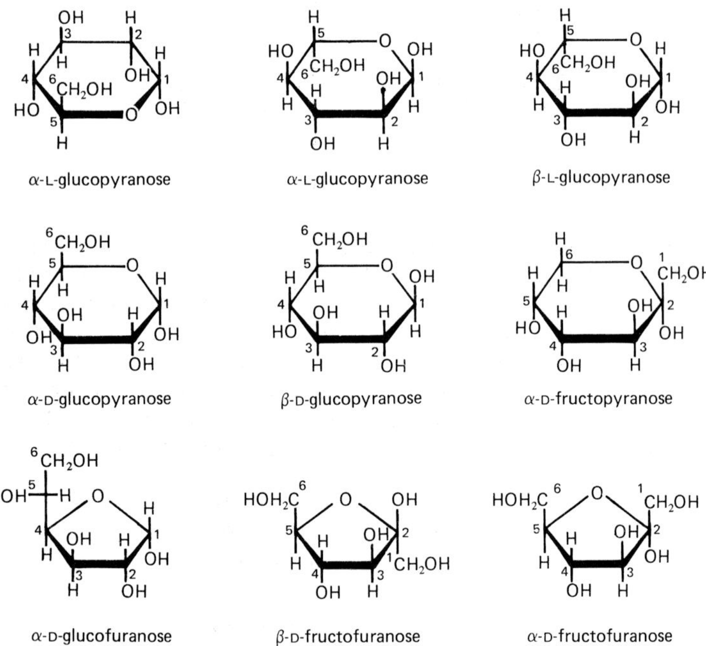

FIG. 2.3. Isomers of glucose and fructose. The obvious shortcomings of the formulae used so far in this chapter as representing molecular structure can be partially resolved, as here, by the adoption of the Haworth convention for ring structures. The ring is treated as planar and drawn to appear perpendicular to the plane of the page. These structures should be compared with those in *Fig. 2.2.* The numbering of carbon atoms may be correlated with that in the open chain structures [2.5] and [2.6] shown on page 4. It should be remembered that only a few of the isomers shown here actually occur naturally in significant amounts and also that not all the possible isomers have been included.

O_2 → H_2O_2 2.14 H_2O → 2.15

The '-uronic' series of sugar acids are aldohexoses with a carboxyl group at carbon 6, such as D-galacturonic acid [2.16] and L-guluronic acid [2.17]. These are important as constituents of polysaccharides such as pectins and alginates but are of little interest in their own right.

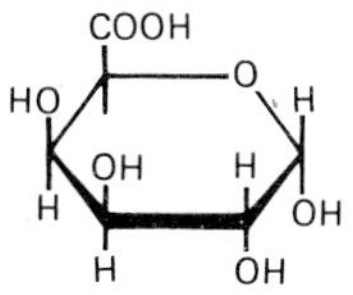

2.16

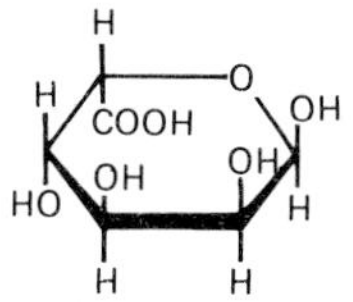

2.17

Reduction of the carbonyl group to a hydroxyl gives sugar alcohols such as xylitol [2.18] and sorbitol [2.19]. Though sweet, these are not absorbed from the intestine. They are synthesised industrially by reduction with hydrogen of the corresponding aldose sugar and are used to replace sugars in diabetic and other 'calorie-reduced' food products. Reduction at other positions gives deoxy sugars such as L-rhamnose (6-deoxy-L-mannose) [2.20], an important minor constituent of pectins, and 2-deoxy-D-ribose [2.21], the sugar component of DNA.

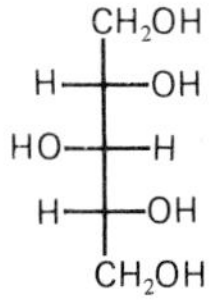

2.18

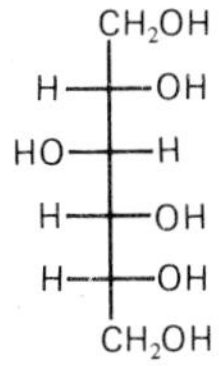

2.19

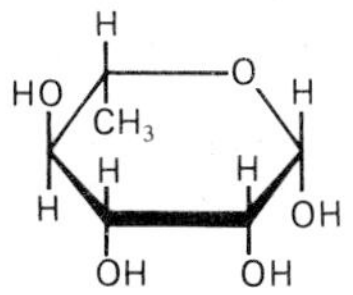

2.20

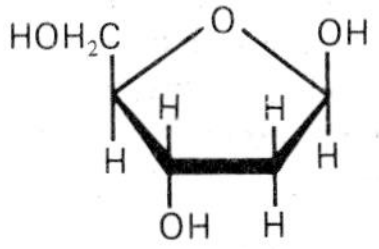

2.21

The most important derivatives of monosaccharides are those in which the 'hemiacetal' or 'reducing' group forms an 'acetal' or 'glycosidic' link with a hydroxyl group of another organic compound:

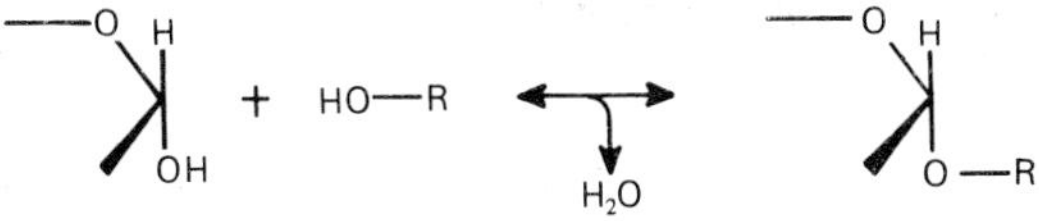

Glycosidic links are stable under ordinary conditions but are readily hydrolysed in acid conditions or in the presence of appropriate hydrolytic enzymes. The formation of the glycosidic link has the effect of fixing the hemiacetal structure in either the α or β configuration and of course abolishing mutarotation. As we will see later in the description of oligosaccharides and other compounds having glycosidic links it will be necessary to specify whether the link in a particular compound is α or β.

Although any compound containing a glycosidic link is, strictly speaking, a glycoside, the term is usually reserved for a class of compounds which occur naturally in plants. These particular glycosides have a sugar component linked to a non-sugar component, termed the aglycone, which is most frequently a flavanoid, or a more simple aromatic substance, or a steroid. The anthocyanin pigments of plants, considered in Chapter 5, are some of the best known glycosides that occur in food but amygdalin [2.22], which occurs particularly in bitter almonds, does give rise to traces of hydrogen cyanide (too little to be hazardous) in some foods.

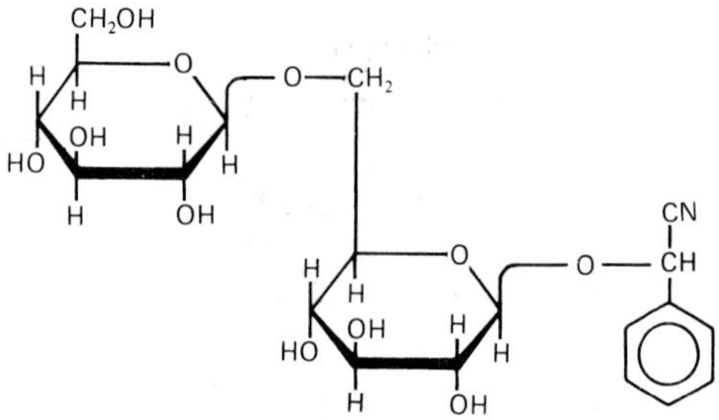

2.22

By far the most numerous and important glycosidic links are those between the reducing groups of one monosaccharide and one of the hydroxyl groups of another. The result is a disaccharide and if the linkage is repeated trisaccharides, tetrasaccharides and, ultimately, polysaccharides will result. The polysaccharides, where hundreds or thousands of monosaccharide units may be combined in a single molecule, are considered in the next section, here we are concerned with the oligosaccharides. (Remember that in Greek 'oligo-' means 'few' whereas 'poly-' means 'many'.) Even when a disaccharide is composed of two identical monosaccharides there are numerous possible structures. This is illustrated in *Fig. 2.4* which shows four of the more important glucose/glucose disaccharides, maltose, cellobiose, gentiobiose and trehalose; many others of course are known. Although the configuration of the hemiacetal involved in the link is no longer free to reverse, it should not be overlooked that the uninvolved hemiacetal (*ie* that of the right hand ring as maltose, cellobiose and gentiobiose are portrayed in *Fig. 2.4*) is still subject to mutarotation in

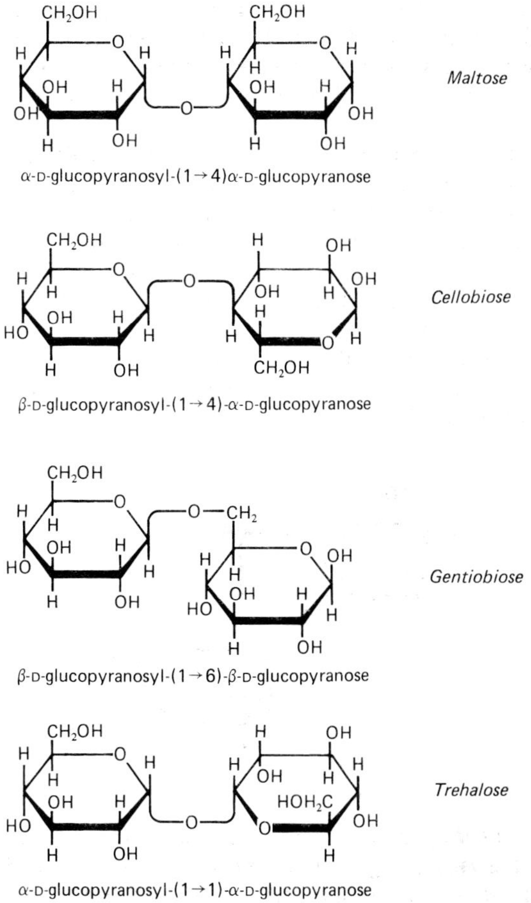

Maltose

α-D-glucopyranosyl-(1→4)α-D-glucopyranose

Cellobiose

β-D-glucopyranosyl-(1→4)-α-D-glucopyranose

Gentiobiose

β-D-glucopyranosyl-(1→6)-β-D-glucopyranose

Trehalose

α-D-glucopyranosyl-(1→1)-α-D-glucopyranose

FIG. 2.4. Disaccharides of glucose. The trivial and systematic names are given. Maltose and cellobiose are breakdown products of starch and cellulose respectively. Gentiobiose is a component of many glycosides including amygdalin. Trehalose occurs in yeast.

---

aqueous solution. Thus these three sugars occur as pairs of α and β anomers. Furthermore, these three sugars continue to show similar reducing properties, associated with the carbonyl group, that monosaccharides have. The failure of trehalose and sucrose to reduce Cu^{2+} ions in alkaline solutions such as 'Fehling's' provides a useful laboratory test to distinguish these sugars from mono- and oligosaccharides which do possess a free reducing group.

The two most important food sugars, lactose and sucrose, are both disaccharides. Lactose, β-D-galactopyranosyl-(1 → 4)-α-D-glucopyranose [2.23], is the sugar of milk (approximately 5 per cent w/v in cow's milk) and is of course a reducing sugar. Sucrose, α-D-glucopyranosyl-(1 → 2)-β-D-fructofuranose [2.24] is the 'sugar' of the kitchen and commerce. The sucrose we buy has been extracted from sugar cane or

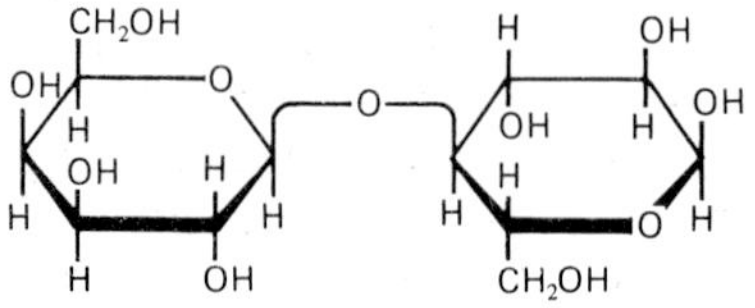

2.23

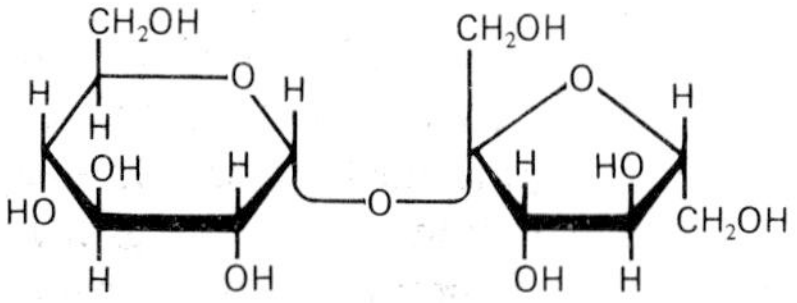

2.24

sugar beet but sucrose is also abundant in most plant materials, particularly fruit. As the glucose and fructose units are joined through both of their hemiacetal groups sucrose is not a reducing sugar. Under mildly acid conditions or the action of the enzyme invertase sucrose is readily hydrolysed to its component monosaccharides. This phenomenon is termed *inversion* and the resulting mixture, *invert sugar*, due to the effect of the hydrolysis on the optical rotation properties of the solution. The specific rotation values for sucrose, glucose and fructose are $+66.5^{\circ}$, $+52.7^{\circ}$ and -92.4° respectively so that we can see that dextrorotatory solution of sucrose will give a laevorotatory solution of invert sugar.

Of the higher oligosaccharides one group deserves particular attention from food chemists. These are the galactose derivatives of sucrose: raffinose (α-D-galactopyranosyl-(1 → 6)-α-D-glucopyranosyl-(1 → 2)-β-D-fructofuranose) and stachyose (α-D-galactopyranosyl-(1 → 6)-α-D-galactopyranosyl-(1 → 6)-α-D-glucopyranosyl-(1 → 2)-β-D-fructofuranose), being the best known. They occur in legume seeds such as peas and beans and present particular problems in the utilization of soya beans. They are neither hydrolysed nor absorbed by the human digestive system so that a meal containing large quantities of beans, for example, becomes a feast for bacteria such as *Escherichia coli* in the large intestine. They produce large quantities of hydrogen and some carbon dioxide as by-products of their metabolism of sugars and the discomforts of flatulence are the inevitable result.

A similar fate befalls the lactose of milk consumed by individuals lacking the enzyme lactase. Humans, like other mammals, have this enzyme as infants but the majority of Orientals and many Negroes lose the enzyme after weaning and are thus unable to consume milk without risk of 'stomach upset'.

Sugars in their crystalline state make important contributions to the appearance and texture of many food products, particularly confectionery, biscuits and cakes. The supermarket shelf usually offers the housewife a choice of two crystal sizes in granulated and caster sugar and a powdered form for icing. The relative proportions of undissolved sugar crystals and saturated sugar solution (syrup) controls the texture of the cream centres of chocolates and similar delights. These considerations apart, the food chemist is primarily interested in sugars when they are in aqueous solution.

Until recently, surprisingly little was known of the behaviour of sugars in solution. Recent studies using nuclear magnetic resonance spectroscopy (NMR) and predictions based on thermodynamic calculations have now revealed a great deal to us. Although the Haworth representation of the configuration of sugar rings always had obvious limitations, after all van't Hoff proposed the tetrahedral arrangement of carbon's valencies in 1874, it is only slowly being superseded in textbooks. The alternative is a 'chair' type structure:

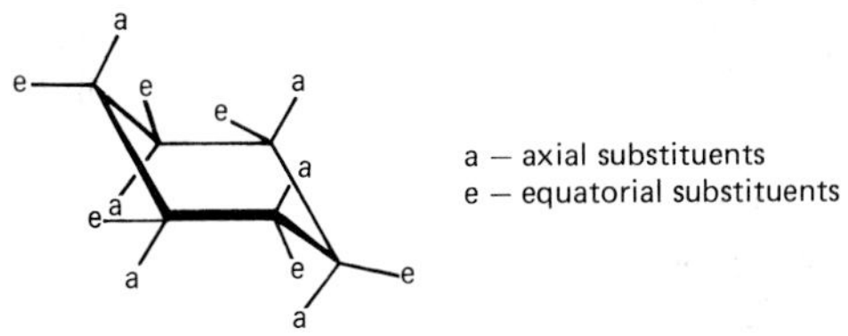
a
e
e
a
e
a
e
a
a
e
e
a
a — axial substituents
e — equatorial substituents

This is the most thermodynamically favourable arrangement for most six-membered aliphatic ring structures. Using β-D-glucose as an example it will be seen that there are in fact two possible chair structures, or conformations (labelled *C*1 and 1*C*), for any single anomer:

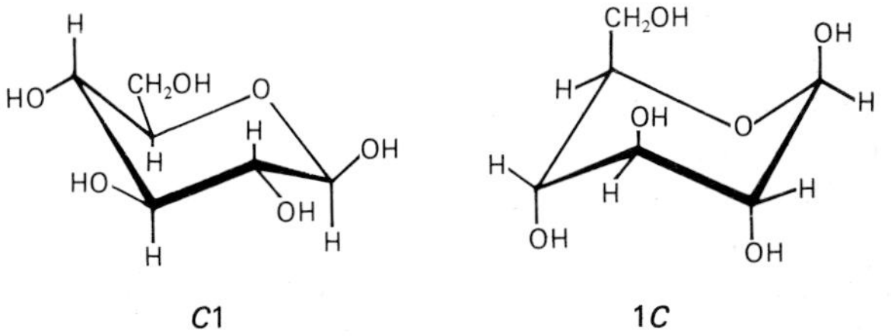
H
HO
CH₂OH
O
H
OH
H
HO
OH
H
H
C1
CH₂OH
OH
H
H
OH
O
H
H
H
OH
OH
1C

Reeves has described a system for naming these different conformations but for most purposes the designation of a particular structure as *C*1 or 1*C* is most easily achieved by comparison with the two structures shown above. If the sugar ring is portrayed as usual with the ring oxygen towards the right and 'at the back' then the conformation is *C*1 when the carbon 1 is below the plane of the 'seat' formed by carbons 2, 3 and 5 and the ring oxygen. The conformation is obviously 1*C* when carbon 1 is above this plane. Theoretical studies have shown that the favoured conformation will be the one which has the greatest number of bulky substituents such as —OH or $-CH_2OH$ in equatorial positions and where neighbouring hydroxyls are as far apart as possible. This has been borne out by experimental work and it immediately becomes obvious why less than 1 per cent of D-glucose molecules in solution are in the 1*C* conformation and why the β anomer is preferred.

The correct description of a furanose ring is more difficult. There are two most probable shapes, described as the envelope [2.25] and twist [2.26] forms. It seems best to assume that furanoses in solution are an

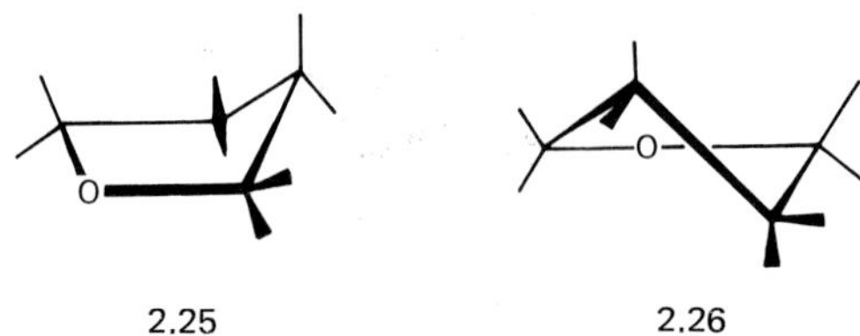

equilibrium of rapidly interconverting forms such as these that do not differ greatly from each other.

Table 2.1. The conformation of the aldohexoses.

	Hydroxyl at			Percentage of			
				Furanose		Pyranose	
	C-2	C-3	C-4	α	β	α	β
D-Allose	eq	ax	eq	5	7	18	70
D-Altrose	ax	ax	eq	20	13	28	39
D-Glucose	eq	eq	eq	<1	<1	36	64
D-Mannose	ax	eq	eq	<1	<1	67	33
D-Gulose	eq	ax	ax	<1	<1	21	79
D-Idose	ax	ax	ax	16	16	31	37
D-Galactose	eq	eq	ax	<1	<1	27	73
D-Talose	ax	eq	ax	<1	<1	58	42
	C3	C4	C5				
1*C*-D-Fructose	eq	eq	ax	<1	25	8	67

The percentages of each anomer and ring size in aqueous solution at 40°C are shown, based on the data of S. J. Angyal (*Angew. Chem. Int. Edn. Engl.*, 1969, 8, 157). The positions of the hydroxyls at carbons 2, 3 and 4 are also shown for the *C*1 conformation. The $-CH_2OH$ group (carbon 6) of a D-aldohexose is always equatorial in the *C*1 conformation, as is the anomeric hydroxyl (on carbon 1) of the β configuration.

An analysis of the spatial distribution of the hydroxyl groups of the D-aldohexoses when in the *C*1 conformation is given in Table 2.1 along with data showing the proportions of the different anomers and ring forms. The *C*1 is the only significant ring conformation in all of the pyranose forms except D-idose and D-altrose. It is clear that with these two sugars no one conformation offers a predominance of equatorial hydroxyl groups. Two factors are seen to control the proportions of the α and β anomers for a particular sugar: the tendency towards an equatorial position for the anomeric hydroxyl has to be balanced against the requirement to be as distant as possible from the hydroxyl on carbon 2. D-fructose occurs predominantly in the 1*C* conformation of the β-pyranose form [2.27] which is seen to have the bulky $-CH_2OH$ group in the equatorial position.

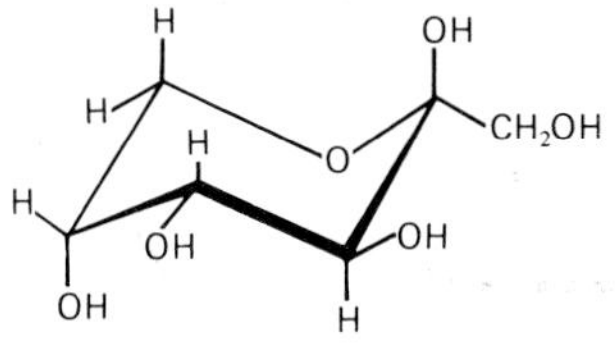

2.27

Perhaps the most striking point to emerge from this type of analysis is the reason for the almost universal role of glucose in living systems, both as a metabolic intermediate and structural element. In spite of the apparent 'untidiness' of its configuration when displayed in the straight chain or Haworth formulae Table 2.1 shows it to be, in its dominant configuration, the one sugar that has *all* its hydroxyl groups in the favoured equatorial position. The conformation and anomer proportions of a particular sugar in solution are not only a function of the intramolecular non-bonding interactions between atoms that have been mentioned so far. They are also dependant on interactions between the sugar molecule and the solvent, usually water in foodstuffs. The fact that there is 37 per cent of the α-anomer of D-glucose at 25 °C in aqueous solution but 45 per cent in pyridine is just one indication of the existence of strong interactions between water and sugar molecules. As we shall see the binding of water to sugars and polysaccharides is an important contributor to the properties of many foodstuffs.

Although the usual crystalline forms of most sugars are anhydrous, and comprise only one particular anomer, D-glucose crystallises from aqueous solutions below 50 °C as the monohydrate of the α pyranose anomer. Likewise, D-lactose crystallises as the monohydrate of the α anomer. Although quite soluble (about 20 g per 100 cm^3 at room temperature) α-D-lactose monohydrate crystals are very slow to dissolve and occasionally form a troublesome gritty deposit in evaporated milk.

The texture of sweetened condensed milk is also highly dependant on the size of the lactose monohydrate crystals.

Not only are sugars generally very soluble in water, they also readily form supersaturated syrups when their solutions are concentrated. The syrups of reducing sugars are frequently very resistant to crystallisation. The confectionery industry takes advantage of this in the production of hard-boiled sweets. These are made by boiling together sucrose, glucose syrup, water and appropriate flavourings and colourings. On cooling the mass sets to a 'glass' or supercooled liquid. The reluctance of glucose syrup to crystallise is also exploited in commercial cake recipes to retain moistness. Very high dissolved sugar concentrations are an essential feature of jam and other 'preserves'.

The effect of dissolved non-electrolytes on the *colligative* properties of solutions, particularly the lowering of the vapour pressure of the solvent, are described by Raoult's Law which states that 'the relative lowering of the vapour pressure of the solvent is equal to the molecular fraction of the solute in solution'. Concentrated sugar solutions do not obey Raoult's Law and the observed vapour pressure can be as much as 10 per cent below the expected figure. Food scientists have avoided the difficulties that would have resulted from this non-ideal behaviour of sugars by the introduction of the concept of water activity. The water activity, a_W, of a solution (or even a solid foodstuff) is equal to the vapour pressure of the solution divided by the vapour pressure of pure water at the same temperature.

The practical importance of controlling water activity in foods will be apparent from the data in Tables 2.2 and 2.3. By lowering the water activity by dehydration, salting or sugaring we can effectively eliminate microbial deterioration. Fortunately, the few bacteria, yeasts and moulds that are adapted to growth at low water activities tend to grow rather slowly and therefore microbial spoilage of low water activity foods is not a common occurrence.

The contrast between water content and activity demonstrated by the data of Table 2.2 should not be overlooked.

Table 2.2. Water content and water activity of various foodstuffs.

	Typical water content (per cent)	Typical water activity (a_W)
Fresh meat	65	0.98
Cheese (Cheddar)	40	0.97
Jam	33	0.88
Salami	30	0.83
Dried fruit	18	0.76
Honey	20	0.75

Table 2.3. Minimum water activities for the growth of micro-organisms.

	a_w
Normal bacteria	0.91
Normal yeasts	0.88
Normal moulds	0.80
Halophilic bacteria	0.75
Xerophilic moulds	0.65
Osmophilic yeasts	0.60

Comparison of the observed vapour pressure of sugar solutions with the higher values expected from Raoult's Law permits a fairly straightforward calculation of the number of water molecules that are no longer effectively part of the bulk of the water of the system, *ie* those that are apparently *bound* to the sugar molecules. The resulting *hydration numbers* approach 2 and 5 molecules of water per molecule of H_2O sugar for glucose and sucrose respectively. As one might expect, the conformation of the sugar molecule affects the hydration number considerably and the fact that the dimensions of the pyranose ring correspond so well with those of the water lattice is reflected in high hydration numbers for sugars compared with the sugar alcohols which do not form rings.

In recent years nuclear magnetic resonance and dielectric relaxation techniques have given us alternative probes for examining water binding. They have revealed a primary hydration layer containing average numbers of water molecules rather higher than indicated by water activity data; for example, ribose 2.5, glucose 3.7, maltose 5.0 and sucrose 6.6. The water that is involved in hydration is not necessarily part of a permanent structure. A water molecule that remains hydrogen-bonded to the sugar molecule for as short a time as one microsecond will be discerned as 'bound water' by NMR techniques. The shorter timescale technique of dielectric relaxation is required to detect more rapid exchange processes. These types of study have also shown that the sugar molecule exerts an influence on the orientation and behaviour of water molecules some way beyond the first layer of this 'hydration shell'.

It is only since we began to understand the behaviour of sugars in solution that any progress could be made towards understanding the most obvious characteristic of sugars in solution – their sweetness. What is frequently overlooked is that most sugars are much less sweet than sucrose and some are not sweet at all. Furthermore, the pursuit of 'slimming' foods has increased our awareness of the fact that sweetness is in no way the prerogative of sugars – saccharin and cyclamate are household words. The measurement of the sweetness of a substance is peculiarly problematical. There are no laboratory instruments to perform the task, no absolute or even arbitrary units of sweetness except 'one lump or two?' in a teacup. Instead we have to rely on the human tongue and the hope that if we average the findings of large numbers of tongues we can obtain useful data. This is one area of biological research where experiments on laboratory animals will never achieve very much. The data we obtain will still not be in absolute units but will be expressed relative to some arbitrary standard, usually sucrose. Table 2.4 gives some relative sweetness data for a number of sugars and other substances. It is a revealing class exercise for students to devise for themselves procedures for comparing the sweetness of say sucrose and glucose and then to compare the results of individual tongues and the 'class average' with data given here.

A vast amount of work has been devoted to identifying the particular molecular feature of sweet tasting substances that is responsible for the sensation of sweetness. One approach is simply to survey as many sweet tasting substances as possible with a view to identifying a common structural element. A second approach has been to prepare large numbers of derivatives of sweet tasting substances in which potentially important groups are blocked or modified. This second approach led Shallenberger's group to propose, in 1967, a general structure for what they termed the 'saporous unit'. They suggested the AH,B system shown in *Fig. 2.5* where A and B represent electronegative atoms (usually oxygen) and AH implies a hydrogen bonding capability.

Table 2.4. The sweetness of sugars and synthetic sweetening agents related to that of sucrose.

Sucrose	1.00
Glucose	0.42
Mannose	0.28
Galactose	0.32
Fructose	0.90
Trehalose	0.55
Lactose	0.16
Sodium cyclamate	30
L-Aspartyl-L-phenylalanine methyl ester (Aspartame)	200
Saccharin	350
Neohesperidin dihydrochalcone	1000
Perillaldehyde antioxime	2000
1-n-propoxy-2-amino-4-nitrobenzene	4000

Data are presented on a molarity basis in the case of the sugars but on a weight basis for the synthetic agents. It should be pointed out that although very sweet, some of these synthetic agents are of dubious value in view of their toxicity and other problems such as stability. (The figures presented here are representative of those quoted by various reserarch groups. Differences in techniques give rise to widely different values being reported from different laboratories.)

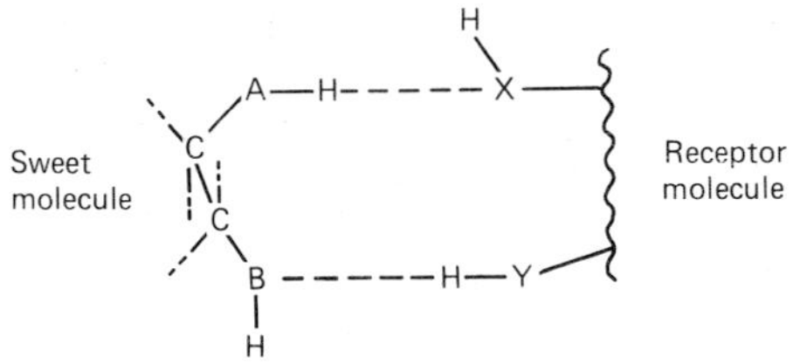

FIG. 2.5. Shallenberger's 'saporous unit'.

For sweetness the distance between B and H needed to be about 0.3 nm. It is assumed that the AH,B system interacts via hydrogen bonding with a similar structure in a sweet-sensitive protein in the epithelium of the taste bud. Such a protein, binding substances in proportion to their sweetness, has actually been isolated from the taste buds of a cow.

Studies of sugar derivatives have shown that the 3,4-α-glycol structure is the primary AH,B system of the aldohexoses. In a ketohexose such as β-D-fructopyranose the anomeric hydroxyl and the hydroxymethylene oxygen have this role.

The conformation of the sugar ring is of crucial importance to the question of sweetness. For the α-glycol AH,B system to have the correct dimensions the two hydroxyl groups must be in the *skewed* (or *gauche*) conformation. If the α-glycol is in the *eclipsed* conformation the two hydroxyl groups are sufficiently close for the formation of an intramolecular hydrogen bond, which excludes the possibility of interaction with a receptor protein. In the *anti* conformation the distance between the hydroxyl groups is too great for interaction with the receptor protein. Inspection of the conformation of *C*1 β-D-glucopyranose (see page 13) shows that all the hydroxyl groups attached directly to the ring (to carbons 1, 2, 3 and 4) are potential 'saporous units'. In contrast 1*C* β-D-glucopyranose has all of its α-glycol structures in the *anti* conformation.

The presence of an AH,B system of the correct dimensions is clearly not the only factor. For example, the *axial* C-4 hydroxyl of D-galactose is able to form an intramolecular hydrogen bond with the ring oxygen thus explaining why D-galactose is less sweet than D-glucose. The fact that the disaccharide trehalose is not twice as sweet as D-glucose tells us that although a sugar may have more than one saporous unit, only one at a time actually binds with the receptor. It should not be surprising that high molecular weight polymers of glucose, such as starch, are quite without taste.

The AH,B concept has been extended to many of the non-sugar sweeteners that were listed in Table 2.4 and some are illustrated in *Fig. 2.6*, together with two sugars. Chloroform earns its place in view

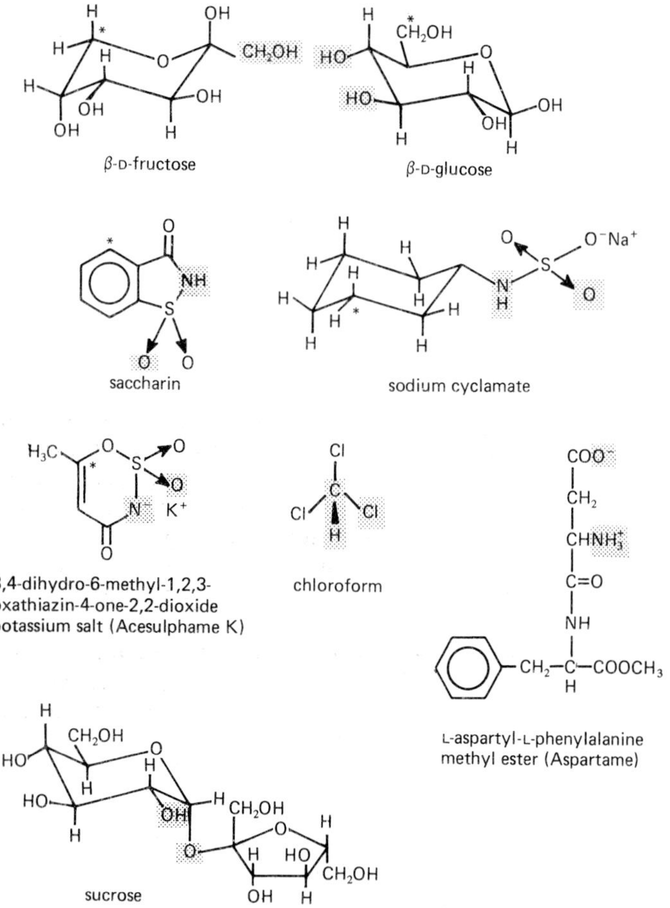

β-D-fructose

β-D-glucose

saccharin

sodium cyclamate

3,4-dihydro-6-methyl-1,2,3-oxathiazin-4-one-2,2-dioxide potassium salt (Acesulphame K)

chloroform

L-aspartyl-L-phenylalanine methyl ester (Aspartame)

sucrose

FIG. 2.6. Saporous units in sweet tasting substances. The atoms presumed to be involved in AH,B systems have been shaded, an asterisk shows the most likely non-polar, or lipophilic, binding site.

---

of its frequent use in the past as 'flavouring' in toothpaste. Detailed studies of aspartame have told us a great deal about the other features of the sweetness receptor site besides the AH,B binding system. The sweetest substances, those which presumably bind tightest to the receptor site, have a non-polar region within the molecule centred about 0.4 nm from atom A of the AH,B system. The nature of this non-polar component does not seem to be critical; L-aspartyl-L-aminomalonyl methyl ester fenchyl ester [2.28] is reported to be some 22 000 times sweeter than sucrose.

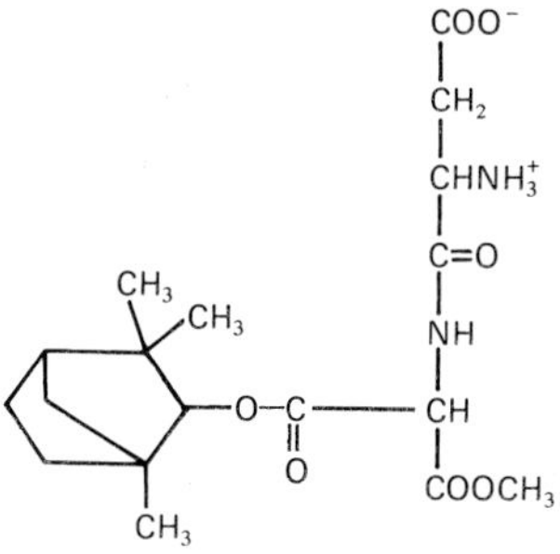

2.28

The fact that most of the sweetest substances we know have their AH,B and non-polar components as part of rigid cyclic structures seems to be a reflection of the inflexibility of the binding site on the receptor protein. An examination of the molecular structures of many antibiotics will provide some thought-provoking comparisons.

Sugars contribute far more to our enjoyment of foods than just sweetness. When they are heated to temperatures above 100 °C a complex series of reactions ensues which gives rise to a wide range of flavour compounds as well as the brown pigments that we associate with caramel and toast. The first stage of the high temperature breakdown is a reversible isomerisation of an aldose sugar to the corresponding ketose via a 1,2-*cis*-enediol intermediate:

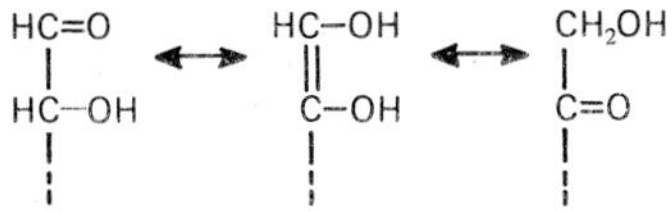

$$\begin{array}{c} HC{=}O \\ | \\ HC{-}OH \\ | \end{array} \longleftrightarrow \begin{array}{c} HC{-}OH \\ \| \\ C{-}OH \\ | \end{array} \longleftrightarrow \begin{array}{c} CH_2OH \\ | \\ C{=}O \\ | \end{array}$$

This reaction (known as the Lobry de Bruyn–Alberda van Eckerstein transformation) requires concerted acid/base catalysis to achieve both the protonation of the carbonyl and the deprotonation of the weakly acidic hydroxyl group. Though at very low concentrations in acidic conditions the anions of organic acids and the hydroxyl ions of water are effective as bases in this system when the temperature is high.

The enediol is readily dehydrated in a neutral or acidic environment via the sequence of reactions shown in *Fig. 2.7*. Hydroxymethylfurfural is by no means the only product of this pathway. Dehydration reactions in the absence of cyclization give a range of highly reactive α-dicarbonyl compounds. Hydroxyacetyl furan [2.29] arises from an alternative cyclization reaction.

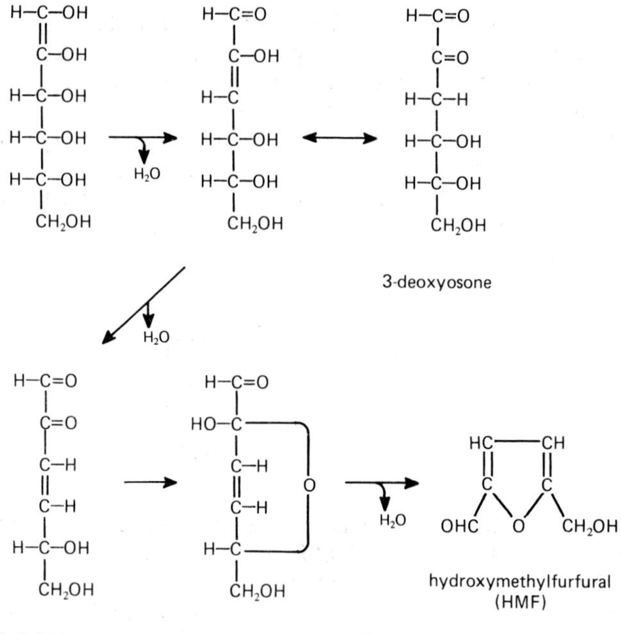

3-deoxyosone

3,4-dideoxyosone

hydroxymethylfurfural (HMF)

FIG. 2.7. The formation of hydroxymethylfurfural by the dehydration of a hexose enediol.

Hydroxymethylfurfural (HMF) can be readily detected in sugar based food products that have been heated such as confectionery, honey which has been adulterated with invert syrup, and 'golden syrup'*. The brown pigments that characterise caramel and other foods arise from a poorly defined group of polymerization reactions. These involve both HMF and its precursors. The pigment molecules have very high relative molecular masses and in consequence are apparently not

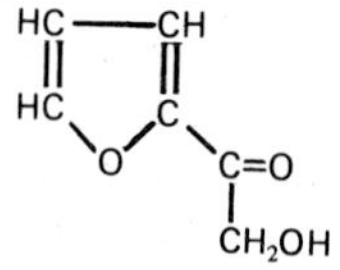

2.29

---

*Commercial invert syrup is produced by the acid hydrolysis of sucrose at high temperatures and is marketed as 'golden syrup'.

absorbed from the intestine. This, together with their traditional application (supposedly safe) to gravy browning *etc* is resulting in their increasing use as a routine brown colouring agent in other foodstuffs.

Small amounts of a vast range of other breakdown products give burnt sugar its characteristic acrid smell; acrolein (propenal, $CH_2{=}CH.CHO$), pyruvaldehyde (2-oxopropanal, $CH_3.CO.CHO$) and glyoxal (ethanedial CHO.CHO) are typical. The caramel flavour itself is reportedly due to two particular cyclic compounds, acetylformoin [2.30] and 4-hydroxy-2,5-dimethyl-3-furanone [2.31].

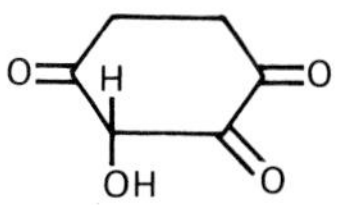

2.30

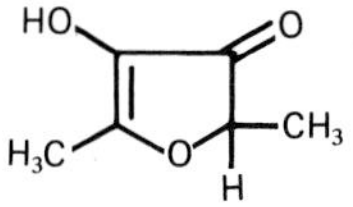

2.31

Many of the carbonyl compounds that are formed during caramelization can be at least tentatively identified by thin layer chromatography of their 2,4-dinitrophenyl hydrazones.*

In the presence of amino compounds the browning of sugars occurs much more rapidly, particularly in neutral or alkaline conditions, in a sequence of events known as the *Maillard reaction*. Low water concentrations, when the reactants are more concentrated, favour the reaction so that it is implicated in the browning of bread crust and the less welcome discoloration that occurs from time to time in the production of powdered forms of egg and milk. These foodstuffs are ones in which sugars are heated in the presence of protein, the source of the amino groups. The amino groups most often involved are those in the side chains of lysine and histidine although the α-amino group of any free amino acids will also participate. The scheme in *Fig. 2.8* shows that the outcome of the Maillard reaction is essentially the same as for the other caramelization reactions in that HMF is produced. However, the *N*-substituted 1-amino-1-deoxy-2-ketoses (the so-called *Amadori compounds*) also give rise to a range of dicarbonyl compounds. Cyclization of these is the source of many of the compounds mentioned earlier such as acetyl formoin [2.30] as well as maltol [2.32] and isomaltol [2.33]. Both of these are important contributors to the flavour of cooked foodstuffs. Disintegration of the dicarbonyl molecules also contributes to the flavour of foodstuffs through the formation of substances such as diacetyl (butanedione, $CH_3.CO.CO.CH_3$), and acetol (hydroxypropanone, $CH_3.CO.CH_2OH$), as well as pyruvaldehyde and glyoxal (ethanedial) that have already been mentioned.

*See: E. F. J. L. Anet, *J. Chromatogr.*, 1962, 9, 291.

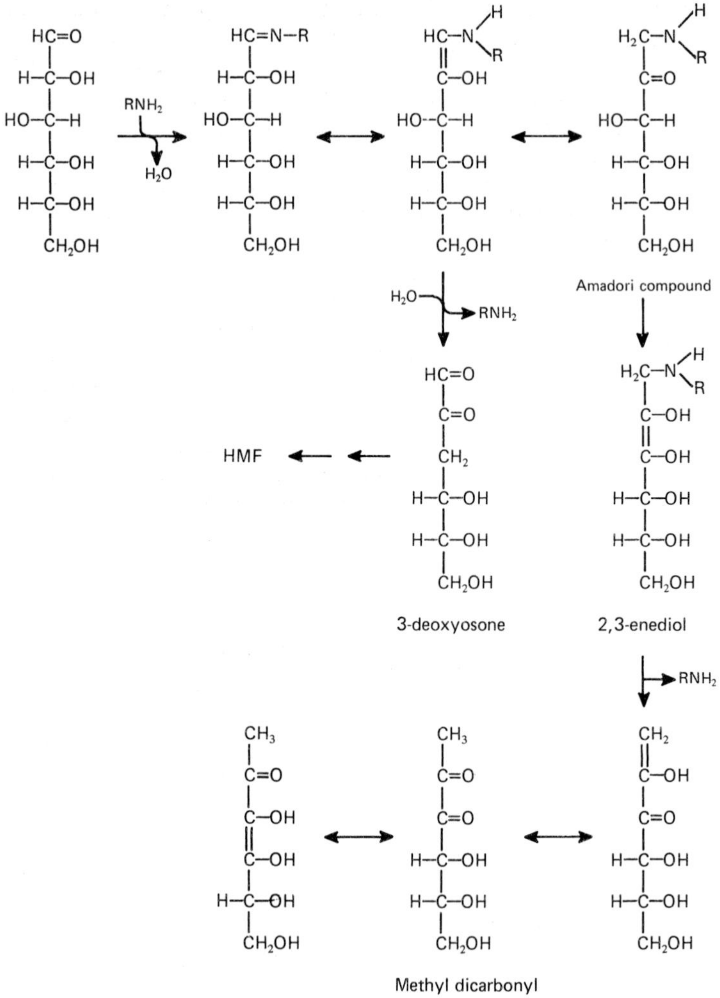

HC=O
H—C—OH
HO—C—H
H—C—OH
H—C—OH
CH_2OH

RNH_2 → H_2O

HC=N—R
H—C—OH
HO—C—H
H—C—OH
H—C—OH
CH_2OH

↔

HC—N(H)(R)
‖
C—OH
HO—C—H
H—C—OH
H—C—OH
CH_2OH

↔

H_2C—N(H)(R)
C=O
HO—C—H
H—C—OH
H—C—OH
CH_2OH

Amadori compound

H_2O → RNH_2

HC=O
C=O
CH_2
H—C—OH
H—C—OH
CH_2OH

HMF ← ←

3-deoxyosone

H_2C—N(H)(R)
C—OH
‖
C—OH
H—C—OH
H—C—OH
CH_2OH

2,3-enediol

→ RNH_2

CH_3
C=O
C—OH
‖
C—OH
H—C—OH
CH_2OH

↔

CH_3
C=O
C=O
H—C—OH
H—C—OH
CH_2OH

↔

CH_2
‖
C—OH
C=O
H—C—OH
H—C—OH
CH_2OH

Methyl dicarbonyl

FIG. 2.8. The formation of HMF and dicarbonyls in the Maillard reaction.

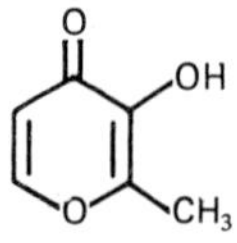

O, OH, O, CH_3

2.32

OH, O, $CO\dot{C}H_3$

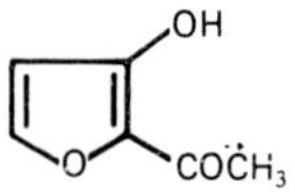

2.33

A most important class of volatile flavour compounds arises by an interaction at elevated temperatures of α-dicarbonyl compounds with α-amino acids known as the *Strecker degradation*. The outcome of the reaction is analogous to enzyme-catalysed transamination but as *Fig. 2.9* shows there is no similarity in the mechanism. The resulting aldehydes make a major contribution to the attractive odour of baking, together with the pyrazine derivatives that result from the dimerization of the sugar residues. Gentle heating of mixtures of a sugar and a range of amino acids will provide a simple laboratory demonstration of the importance of these reactions.

While most of the effects of the Maillard reaction are regarded as favourable there can be adverse nutritional consequences. When foodstuffs containing protein and carbohydrates and having a low water content (such as cereals or milk powder) are held at even quite modest temperatures the Maillard reaction will take place and cause losses of amino acids. The amino acids involved will be those which have a free amino group in their side chain. Lysine is the most reactive of these, followed by arginine and then tryptophan and histidine.

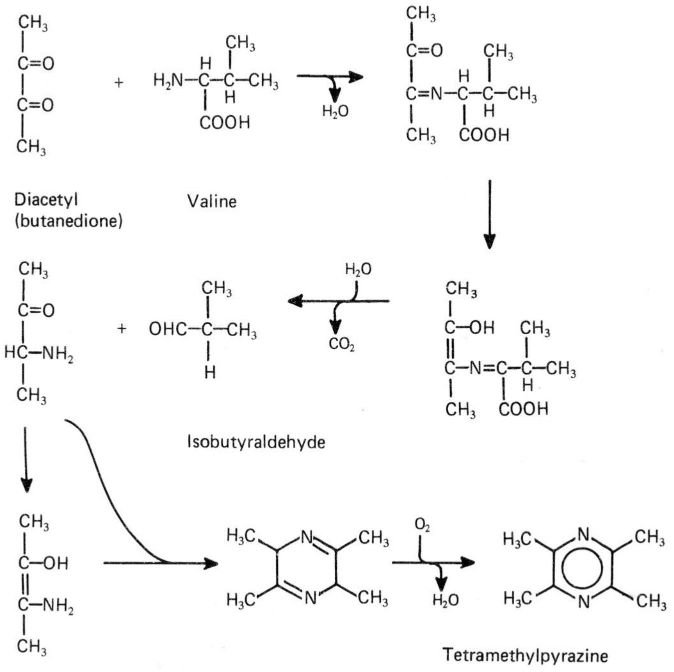

FIG. 2.9. The Strecker degradation reaction between diacetyl (butanedione) and valine.

At temperatures as low as 37 °C a mixture of glucose and casein (the principal protein of milk) may lose as much as 70 per cent of its lysine over a period of five days. Under these conditions the reaction does not go beyond the formation of the Amadori compound so that no visible browning occurs. The amino acid in this form has no nutritional value as it is not susceptible to the normal processes of digestion and absorption in the intestine. If large numbers of the amino acid side chains of a protein molecule are carrying sugar residues as a result of the Maillard reaction then the entire molecule may become unavailable due to blocking of the proteolytic enzymes of the intestine. It is when one realises that lysine, histidine and tryptophan are all essential amino acids, *ie* they must be supplied in the diet (see Chapter 4) and that in poor vegetarian diets there can often be a deficiency of lysine that the importance of retarding the Maillard reaction in stored foods becomes most obvious.

Polysaccharides

In the first part of this chapter we considered the properties of the mono- and oligo-saccharides. We can now turn to the high molecular weight polymers of the monosaccharides – the polysaccharides. The considerable diversity of polysaccharide structures are often classified according to features of their chemical structure. For example, are the polymer chains branched or linear, is there more than one type of monosaccharide residue present? Unfortunately, these classifications have little or no relationship to either the natural role of a polysaccharide or its behaviour in food systems. The only common feature of the polysaccharides that are important to food scientists is that they occur naturally in plants. Polysaccharides play a relatively minor role in animals. Glycogen, which is very similar structurally to amylopectin, acts as an energy/carbohydrate reserve in the liver and muscles. The other animal polysaccharides are chitin, a polymer of *N*-acetylglucosamine [2.34], the fibrous component of arthropod exoskeletons, and the proteoglycans and mucopolysaccharides which have specialised roles in connective tissues. The molecules of these connective tissue polysaccharides include considerable proportions of protein.

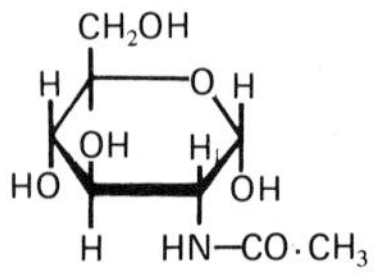

CH_2OH

H O H

OH H

HO OH

H HN—CO·CH_3

2.34

Polysaccharides have two major roles to play in plants. The first, as an energy/carbohydrate reserve in tissues such as seeds and tubers, is almost always filled by starch. The second, providing the structural skeleton of both individual plant cells and the plant as a whole, is filled by a wide range of different structural types which offer a correspondingly wide range of physical characteristics. The contribution of polysaccharides to food is no less diverse. From a strictly nutritional point of view starch is the only one of the major plant polysaccharides that is readily digested in the human intestine and thereby utilized as a source of carbohydrate. In fact a high proportion of human requirements for energy is met by the starch of cereal grains and tubers such as potatoes and cassava. Only in recent years has the nutritional significance of the indigestible polysaccharides in our diet become apparent. Cellulose, pectic substances and hemicelluloses, often known collectively as *fibre*, have an essential role in the healthy function of the large intestine.

In spite of their physiological importance it is through their influence on the texture of food that polysaccharides often make the most immediate impact on the consumer. Furthermore, the relationship between a food's texture and the underlying molecular structure of its components is being revealed most clearly in the case of polysaccharides and this relationship therefore provides the theme of the accounts of the more important food polysaccharides that follow.

Starch

Reflecting its role as a carbohydrate reserve, starch is found in greatest abundance in plant tissues such as tubers and the endosperm of seeds. It occurs in the form of granules which are usually an irregular rounded shape and ranging in size from 2 to 100 μm. Both the shapes and sizes of the granules are characteristic of the species of plant and can be used to identify the origin of a starch or flour.

Starch consists of two types of glucose polymer, *amylose*, which is essentially linear, and *amylopectin*, which is highly branched. They occur together in the granules but amylose may be separated from starch solutions since it is much less soluble in organic solvents such as butanol. Most starches are 20–25 per cent amylose but there are exceptions: pea starch is around 60 per cent amylose and the so-called 'waxy' varieties of maize and other cereals have little or none.

Amylose consists of long chains of α-D-glucopyranosyl residues linked, as in maltose, between their 1 and 4 positions (see *Fig. 2.10*). There is no certainty about the length of the chains but they are generally believed to contain many thousands of glucose units so that typical, average, molecular weights are between 2×10^5 and 10^6. It is becoming clear that amylose chains do contain a very small amount of branching of the type which is characteristic of amylopectin.

Amylopectin is a much larger molecule, having about 10^6 glucose units per molecule. As in amylose the glucose units are joined by α1–4 glycosidic links but some 4–5 per cent of the glucose units are also in-

(*i*)

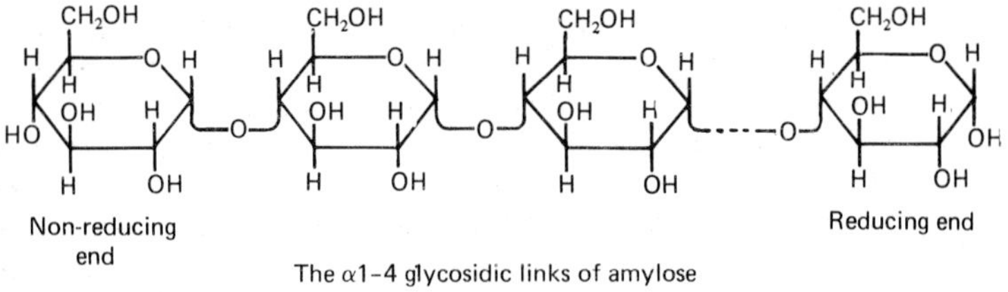

Non-reducing end

Reducing end

The α1–4 glycosidic links of amylose

(*ii*)

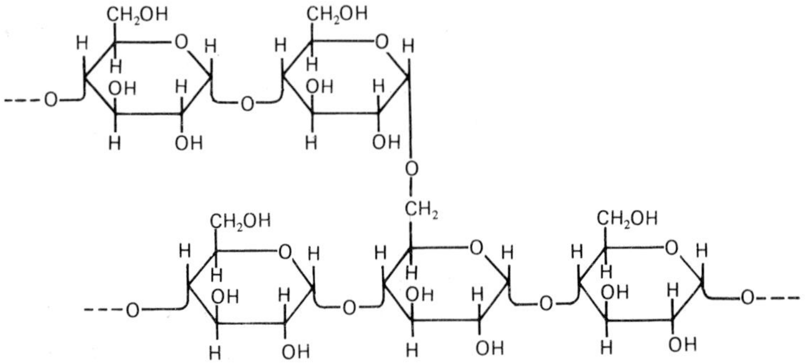

The α1–6 branch of amylopectin

FIG. 2.10. Amylose and amylopectin.

volved in α1–6 links creating branch points as shown in *Fig. 2.10*. This proportion of branch points results in an average chain length of 20-25 units. Over the years a great many hypotheses have been put forward to describe the arrangement of the branching chains of amylopectin and for many years the tree like arrangement first proposed by Meyer and Bernfeld in 1940 was accepted. However, the availability of the enzyme, pullulanase, which is specific for α1–6 linkages, and gel filtration methods for the determination of the chain lengths of the resulting fractions, has given us new insights. *Figure 2.11* shows one possible arrangement of the chains which is gaining wide acceptance. Its essential feature is a skeleton of singly branched chains approximately 25 glucose units long which carry clusters of mostly unbranched chains approximately 15 units long.

Surprisingly little is known of the arrangement of the amylose and amylopectin molecules within the starch granule. When observed in the polarizing microscope they show the 'Maltese Cross' pattern which is characteristic of birefringent materials. This implies that there is a high degree of molecular orientation but gives us no details. The crystallinity is confirmed by the X-ray diffraction patterns which incidentally reveal that, in general, root starch granules are more crystalline than those of cereals. An unexpected aspect of the crystallinity is that it is owed to the branched amylopectin rather than the linear amylose. This is indicated by the observation that waxy starches, which lack amylose, give an X-ray pattern very similar to that of a normal starch. It seems most likely that the clusters of amylopectin short chains (see *Fig. 2.11*) are the crystalline regions of the molecule. Furthermore, these clusters could well be the origin of the concentric layers of crystalline, relatively acid-resistant material, that have been observed in sectioned granules.

Undamaged starch granules are insoluble in cold water due to the collective strength of the hydrogen bonds binding the chains together, but as the temperature is raised to what is known as the *initial gelatinisation temperature* water begins to be imbibed. The initial gelatinisation

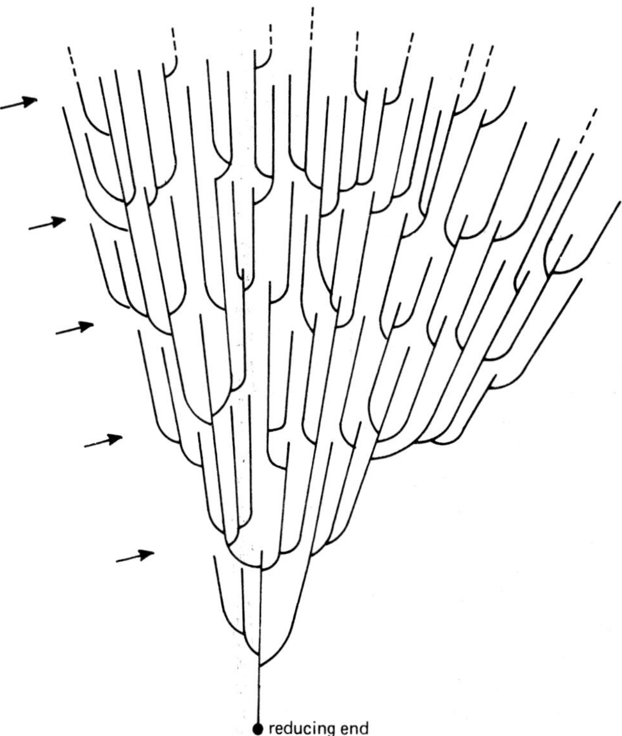

FIG. 2.11. An amylopectin structure. This suggested arrangement of the branched amylopectin chains is based on Robin's studies of potato starch. It is generally assumed that the amylopectin chains will be orientated radially within the starch granule presenting their non-reducing ends towards the surface. The arrows mark concentric regions, approximately 6 nm apart, where branch points are concentrated.

temperatures are characteristic of particular starches but usually lie in the range 55–70 °C. As water is imbibed the granules swell and there is a steady loss of birefringence. Studies with X-ray diffraction show that complete conversion to the amorphous state does not occur until temperatures around 100 °C are reached. As swelling proceeds and the swollen granules begin to impinge on each other the viscosity of the suspension/solution rises dramatically. The amylose molecules are leached out of the swollen granules and also contribute to the viscosity of what is best described as a paste. If heating is maintained together with stirring, the viscosity soon begins to fall as the integrity of the granules is lost. If the paste is allowed to cool the viscosity rises again as hydrogen bonding relationships between both amylopectin and amylose are re-established to give a more gel-like consistency. These changes will be familiar to anyone who has used starch to thicken gravy or prepare desserts such as blancmange. These changes can be followed in the laboratory with viscometers such as the Brabender Amylograph which apply continuous and controlled stirring to a starch suspension during a reproducible temperature regime. Some typical Brabender Amylograph results for a number of different starches are shown in *Fig. 2.12*. Although our understanding is increasing we are still some way from being able to explain all the distinctive features of a particular Brabender trace in terms of the molecular characteristics of the starch.

Solutions of starch have a number of characteristic properties of which the reaction with iodine is the best known. Solutions of iodine

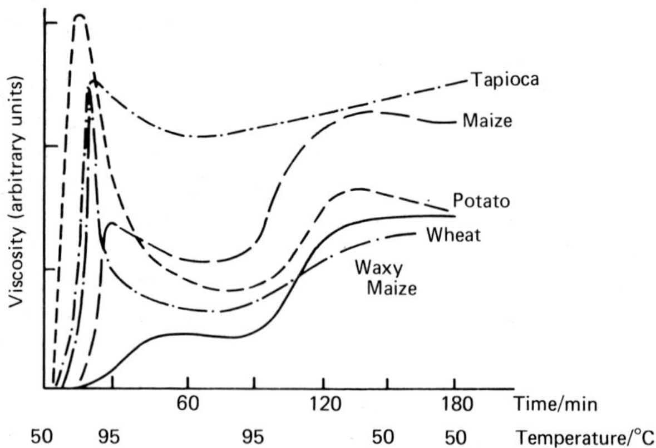

FIG. 2.12. Brabender Amylograph traces of the gelatinisation of a number of starches. The temperature is raised steadily from 50 °C to 95 °C over the first half hour period. It is then maintained at 95 °C for one hour and finally brought back down to 50 °C over a further half hour period. (*Dr D. Howling and Applied Science Publishers.*)

in aqueous potassium iodide stain the surface of starch granules dark blue but the reaction is most apparent with the amylose component of dissolved starch. In the presence of iodine the amylose chains are stabilized in helical configuration with six glucose residues to each turn of the helix and the iodine molecules complexed down its axis. The blue colour of the complex is dependant on the length of the chain involved; with 100 residues λ_{max} is at 700 nm but with shorter chains λ_{max} falls until with only 25 residues it is down to 550 nm. In view of this it is not surprising that amylopectin and glycogen both give a reddish brown colour with iodine.

When starch solutions or pastes are allowed to stand for a few hours they begin to show changes in their rheological properties. Dilute solutions lose viscosity but concentrated pastes and gels become rubbery and exude water. Both types of change are due to a phenomenon termed *retrogradation* which involves just the amylose molecules. Over a period of time these associate together and effectively 'crystallise out'. If one recalls that the role of amylose in starch solutions is to bind together the expanded granule structure of amylopectin molecules, then these effects of retrogradation on their rheological properties are obviously inevitable.

Retrogradation is the cause of a number of undesirable changes in food products. For example, starch is frequently used to thicken the juice or gravy of commercial fruit or meat pies. If, after manufacture the pie is frozen the amylose will undergo retrogradation rapidly so that on thawing the paste will be completely liquid, revealing just how much, or how little, of the filling was actually meat or fruit. The answer to this has been the use of starch from the 'waxy' varieties of maize which contain only amylopectin and therefore give pastes that will survive freezing and thawing.

In the manufacture of bread and other bakery products the behaviour of starch is obviously very important. When a dough is first mixed the damaged starch granules (the damage is an inevitable consequence of flour milling) absorb some water. During breadmaking the kneading and proving stages prior to the actual baking allow time for the α- and β-amylases that are naturally present in flour to break down a small proportion of the starch to maltose and other sugars. These are fermented by the yeast to give the CO_2 which leavens the dough. Other baked goods rely on the CO_2 from baking powder together with air, whipped into the mixture, for the same purpose.

Once in the oven the starch granules gelatinise and undergo varying degrees of disruption and dispersion. The α-amylase activity of flour persists until temperatures around 75 °C are reached so that some enzymic fragmentation of starch granules also occurs, particularly in bread. Amongst the factors that affect the degree of starch breakdown are availability of water and the presence of fat. Obviously, in a shortbread mixture containing only flour, butter and sugar the coating of fat over the starch granules will limit the access of the small amount of water present and we find that, after baking, the granule structure is still perceptible. (A *roux* is another example where fat is used to control the gelatinisation behaviour of starch.)

After one or two days cakes with a low fat content, and bread, start to go stale. Staling is usually thought of as a simple drying out process but of course it occurs just as readily in a closed container. What is happening is that the starch is undergoing retrogradation. The crystallinity of the retrograded amylose gives the crumb of stale bread its extra 'whiteness' and the increased rigidity of the gelatinised starch causes the lack of 'spring' in the crumb texture. Retrograded amylose can be returned to solution only by heating, *ie* regelatinisation; stale bread is often 'revived' by moistening it and then returning it to a hot oven for a few minutes.

Pectins

Pectins are the major constituents of the middle lamella of plant tissues and also occur in the primary cell wall. They comprise a substantial proportion of the structural material of soft tissues such as the parenchyma of soft fruit and fleshy roots. There is no certainty as to the exact nature of pectins as they occur in plant tissues – the term protopectin is applied to water insoluble material that yields water soluble pectin on treatment with weak acid. Most elementary accounts of pectins describe them as linear polymers of α-D-galacturonic acid linked through the 1 and 4 positions, with a proportion of the carboxyl groups esterified with methanol as shown in *Fig. 2.13*. Their molecular weights can best be described as uncertain but high, probably around 100 000. The degree of esterification in unmodified pectins, *ie* those that have not been deliberately de-esterified during extraction or processing, can vary from around 60 per cent, *eg* those from apple pulp and citrus peel, to around 10 per cent, *eg* that from strawberry.

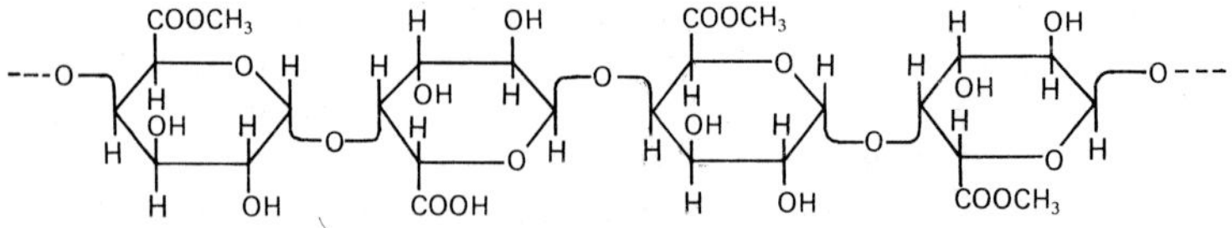

FIG. 2.13. The polygalacturonic acid of pectin.

Although this simple structure was always recognised by carbohydrate chemists to be an over-simplification it is only in recent years that food scientists have begun to appreciate that the behaviour of pectins in food systems such as jam and confectionery could be properly understood only if minor components and structural features were also taken into account. As many as 20 per cent (but usually rather less) of the sugars in a pectin may be found to be neutral sugars such as L-rhamnose, L-arabinose, D-glucose, D-galactose and D-xylose. L-rhamnose residues are found in all pectins, bound by a 1,2- link into the main chains. The other sugars apparently occur in short branches linked mostly to the L-rhamnose residues but also to the D-galacturonate residues of backbone. The detailed structure of these side chains, and the mode of their attachments, remain to be fully elucidated but the structural features shown in *Fig. 2.14* are likely to prove a fairly

$$---\rightarrow 4\text{GalA}\alpha 1 \rightarrow 2\text{Rha}\beta 1 \,[\rightarrow 4\text{GalA}\alpha 1]_m\!\rightarrow 2\underset{\underset{\text{A/G}}{\uparrow}}{\text{Rha}}\beta 1 \rightarrow 4\text{GalA}\alpha 1 ---$$

$$---\rightarrow 2\underset{\underset{\text{A/G}}{\uparrow}}{\text{Rha}}\beta 1 \rightarrow 4\text{GalA}\alpha 1 \rightarrow 4\underset{\underset{\text{X}}{\uparrow}}{\text{GalA}}\alpha 1 \,[\rightarrow 4\text{GalA}\alpha 1]_n\!\rightarrow$$

FIG. 2.14. A probable structure for a typical pectin. Note: (*i*) GalA indicates D-galacturonate, with or without methylation and/or acetylation. (*ii*) Rha indicates L-rhamnose. (*iii*) 'α1 → 4', 'α1 → 2' and 'β1 → 4' indicate the type of glycosidic link. (*iv*) The lengths of the homogalacturonan chains (*n* and *m*) are likely to lie between 20 and 50. (*v*) A/G and X are, respectively, arabinogalactan and xylan side chains.

accurate representation of a typical pectin. A further complication is that some of the hydroxyl groups on C-2 and C-3 of the D-galacturonate residues are often acetylated. In summary we can describe the pectin chain as having fairly long 'smooth' regions of galacturonan* interspersed with short 'hairy' regions of side chains and kinks induced by the presence of L-rhamnose (see *Fig. 2.15*).

The importance of pectin in food is its ability to form the gels that are the basis of jam and other fruit preserves. Gels consist largely of water and yet are stable and retain the shape given to them by a mould. We have already seen that sugar molecules, and by inference those of polysaccharides also, are effective in binding water. However, a stable gel also requires a stable three-dimensional network of polymer chains with water, together with dissolved substances and solid particles, held in its interstices. Until recently it was frequently assumed that simple interactions between neighbouring chains, such as single hydrogen bonds, were all that was required to maintain the stability of the network. The fact that the lifetime of an isolated hydrogen bond in an aqueous environment is only a tiny fraction of a second was quite overlooked. It is obvious that attractive forces of this character provide an adequate explanation for the viscosity of the solutions of many

*Terms such as galacturon*an*, gluc*an* and arabinogalact*an* describe polysaccharide chains comprising, respectively, galacturonic acid, glucose and arabinose together with galactose.

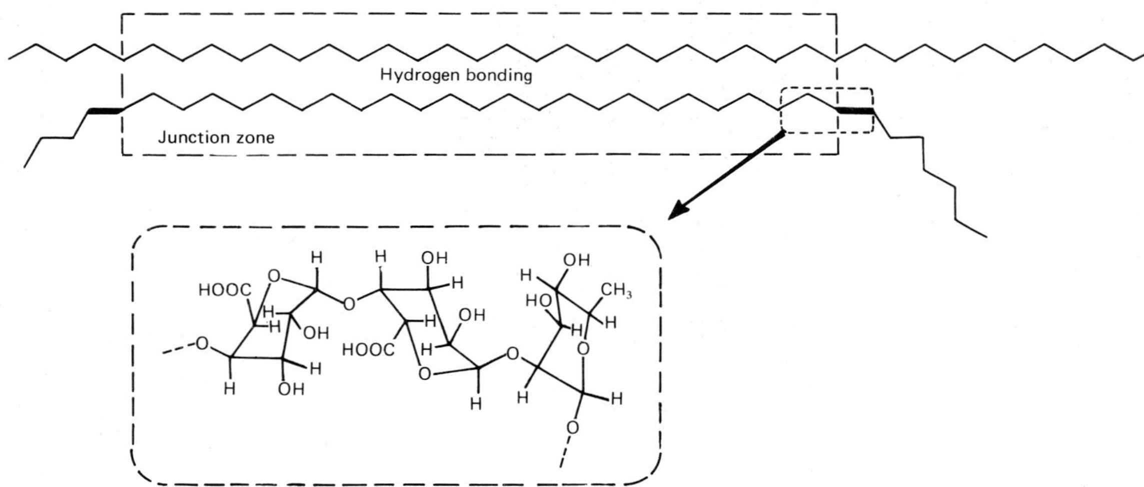

FIG. 2.15. A pectin junction-zone. Two galacturonan chains are bound together by hydrogen bonding in a region of ordered structure at least 20 units in length and terminated by L-rhamnose residues in one or other of the chains. The configurations of the glycosidic links involving L-rhamnose are not finally established but the most probable are shown here.

polymers, such as the polysaccharide gums, but do not explain gel structure.

The concept of the *junction-zone* was introduced by Rees to explain the properties of carageenan gels. Carageenans are members of a group of polysaccharides based on sulphate derivatives of L and D galactose (agar, of microbiological fame, is another member of the group) which occur in one family of seaweeds, the *Rhodophyceae*. Although single hydrogen bonds are not sufficiently durable to provide a stable network a multiplicity of hydrogen bonds or other weak forces, acting cooperatively, are the basis of the stability of crystal structures. Rees has shown that polymer chains in a gel interact in regions of ordered structure involving considerable lengths of the chains. These regions, the junction-zones, are essentially crystalline in nature and details of their structure have been revealed by X-ray crystallography.

The detailed structure of the junction-zone is dependant on the nature of the polymer involved. In agar and carageenan gels the interacting chains form double helices which further interact to form 'super-junctions'. The pectins have been less studied but it is clear that the two interacting chains pack closely together as suggested in *Fig. 2.15*. It is now possible to understand the importance to gel structure of the so-called anomalous residues, such as L-rhamnose, and the 'hairy' region of the chain. Without them there would be nothing to prevent interaction taking place along the entire length of the chains resulting in precipitation. This is precisely the phenomenon that is involved when the unbranched, uninterrupted chains of amylose undergo retrogradation. The generalised two-dimensional picture of a polysaccharide gel that emerges from these observations is presented in *Fig. 2.16*.

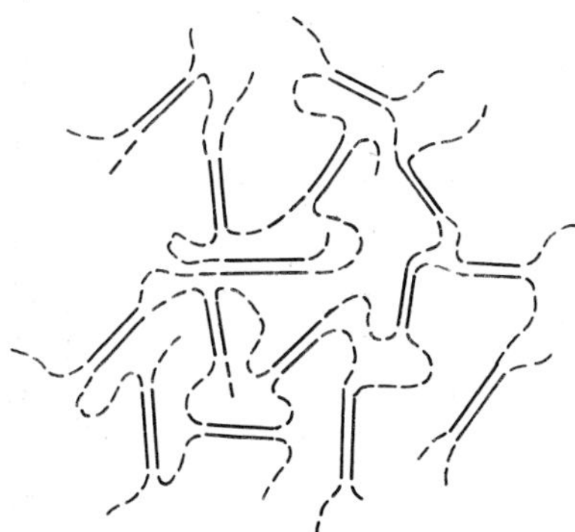

FIG. 2.16. A generalised, two-dimensional, view of a polysaccharide gel. Regions of the polymer chain involved in junction zones are shown: ═══. Other regions of irregular orientation, so-called random coils, are shown: - - - - -. Water, dissolved substances and suspended particles occupy the spaces between the polymer chains. Each end of the junction zone is terminated by an anomalous residue or the presence of a bulky side chain on at least one of the chains involved.

It may have been inferred from the discussion so far that pectins form gels quite readily but any one who has been involved in jam-making at home will know how untrue this is. The presence of pectin, solubilised by boiling the fruit, is itself, not enough. For successful jam-making we must meet other conditions are well. A highly esterified pectin (at least 70 per cent of the carboxyl groups methylated) is advisable with a concentration of at least 1 per cent w/v. The acid present, from the fruit or added, must maintain the pH below 3.5. The total sugar* content must be at least 50 per cent w/v. While the relationships between these requirements and the procedures of jam-making can be readily discerned the reasons for them are less obvious. In fact pectin is a reluctant participant in gel formation. The low pH and high degree of esterification ensure that very few galacturonate residues are in the ionized (*ie* $-COO^-$) form so that the affinity of the polymer chains for water is minimized. The high sugar content effectively competes for water with the polymer chains with the result that the amount of bound water which would otherwise disrupt the fragile affinities of the junction-zone is reduced and the gel can form.

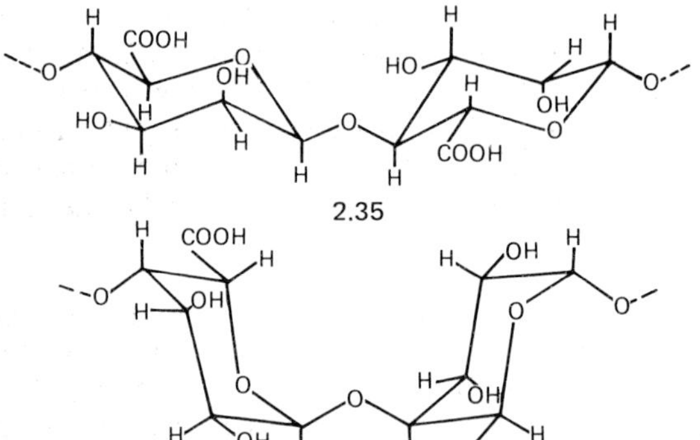

2.35

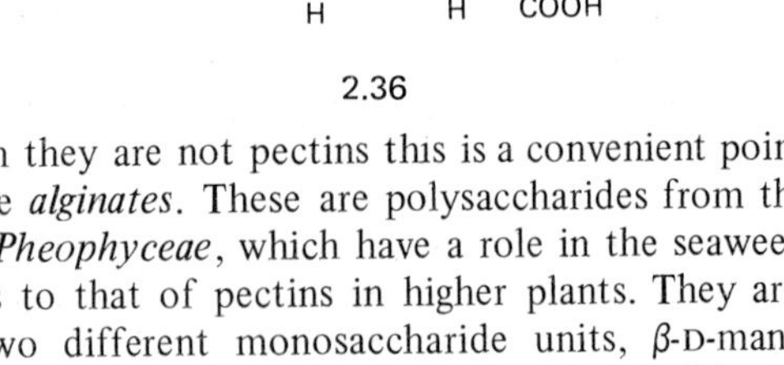

2.36

Although they are not pectins this is a convenient point at which to mention the *alginates*. These are polysaccharides from the brown seaweeds, the *Pheophyceae*, which have a role in the seaweed thallus that corresponds to that of pectins in higher plants. They are linear polymers of two different monosaccharide units, β-D-mannuronic acid [2.35] and α-L-guluronic acid [2.36]. These two will be seen to be C-6 epimers of each other. Both occur together in the same chains but linked in three different types of sequence:

—M—M—M—M—, —G—G—G—G—G—, —M—G—M—G—M—G—,

---

*A proportion of the sucrose added is hydrolysed to glucose and fructose during the boiling.

in each case by 1–4 links. It will be noticed that the pyranose rings of these two sugar acids take up the conformation that places the bulky carboxyl group in the equatorial position.

Alginic acid itself is insoluble but the alkali metal salts are freely soluble in water. Gels readily form in the presence of Ca^{++} ions. Rees' group has shown that the junction zones only involve —G—G—G—G— sequences in what is described as an 'egg box' structure with a Ca^{++} ion complexed by four L-guluronate residues:

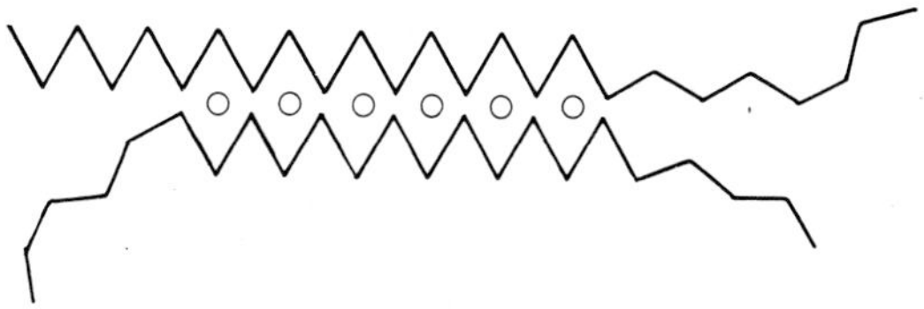

Ca-alginate gels do not melt below the boiling point of water and they have therefore found a number of applications in food products. If a fruit puree is mixed with sodium alginate and is then treated with a calcium-containing solution various forms of reconstituted fruit can be obtained. For example, if a cherry/alginate puree is added as large droplets to a calcium solution one will obtain the convincing synthetic cherries popular in the bakery trade. Rapid mixing of a fruit/alginate puree with a calcium solution followed by appropriate moulding will give the 'pieces' of apple or apricot that are now sold as pie fillings.

Cellulose and hemicelluloses

To most people the terms 'fibre' or 'roughage' indicate the fraction of plant foods, particularly cereals, that are not digested in the human alimentary tract. To the food analyst 'fibre' is the insoluble residue that remains after a defatted foodstuff has been extracted first with boiling dilute acid and then with boiling dilute alkali, a treatment that solubilises almost all but the cellulose and lignin of plant tissues. Once again definitions have proved to be of very little value to us.

Carbohydrate chemists group the indigestible (*ie* by humans) polysaccharides of plants into thee classes, the pectins that we have already discussed, cellulose and the hemicelluloses. In the consideration of the possible benefits of high fibre diets it is not usual to distinguish between these classes in spite of their lack of common chemical features.

Cellulose, said to be the most abundant organic chemical on earth, is an essential component of all plant cell walls. It consists of linear molecules of at least 3000 β1–4 linked glucopyranose units. This linkage leads to a flat ribbon arrangement maintained by intramolecular hydrogen bonding as shown in *Fig. 2.17.* Within plant tissues cellulose molecules are aligned together to form *microfibrils* several micro-

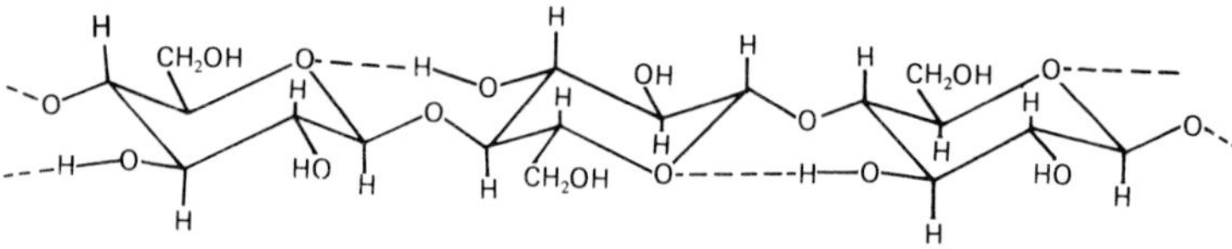

FIG. 2.17. The flat-ribbon arrangement of the cellulose chain. Broken lines indicate hydrogen bonds.

metres in length and about 10–20 nm in diameter. Although doubt still exists as to the exact arrangement of the molecules within the microfibrils it is clear that the chains pack together in a highly ordered manner maintained by intermolecular hydrogen bonding. It is the stability of this ordered structure that gives cellulose its insolubility in virtually all reagents and also the great strength of the microfibrils. Herbivorous animals, particularly the ruminants, are able to utilize cellulose by means of the specialised micro-organisms that occupy sections of their alimentary tracts. These secrete cellulolytic enzymes that release glucose and then they obtain energy by fermenting the glucose to short chain fatty acids, such as butyric acid, which can then be absorbed and utilized by the animal. Even in these specialised animals, cellulose digestion is a slow process as evidenced by the disproportionately large abdomens of herbivores compared with carnivores.

The hemicelluloses were originally assumed to be low molecular weight (and therefore more soluble) precursors of cellulose. Although this idea has long been discarded we are left with a most unfortunate name to describe this group of structural polysaccharides. They are closely associated with the cellulose of plant cell walls from which they can be extracted with alkaline solutions (*eg* 15 per cent KOH). Three types of hemicellulose are recognised, the xylans, the mannans and glucomannans, and the galactans and arabinogalactans. Although the leaves and stems of vegetables provide a high proportion of the indigestible polysaccharide of our diet very little is known about the hemicelluloses of such tissues. We know much more about the hemicelluloses of cereal grains where the xylans are the dominant group. The essential feature of these xylans is a linear or occasionally branched backbone of β1–4 linked xylopyranose residues with single L-arabinofuranose and D-glucopyranosyluronic acid residues attached to the chain as indicated in *Fig. 2.18*. Xylans are major constituents of the seed coats of cereal grains which on milling constitute the bran. A typical wheat bran composition (on a dry basis) is lignin* 8 per cent, cellulose 30 per cent, hemicellulose 25 per cent, starch 10 per cent, sugars 5 per cent, protein

*Lignin is the characteristic polymer of woody tissues. Although the detailed structure is still not fully understood it is based on the condensation of aromatic aldehydes such as vanillin, syringaldehyde and p-hydroxybenzaldehyde.

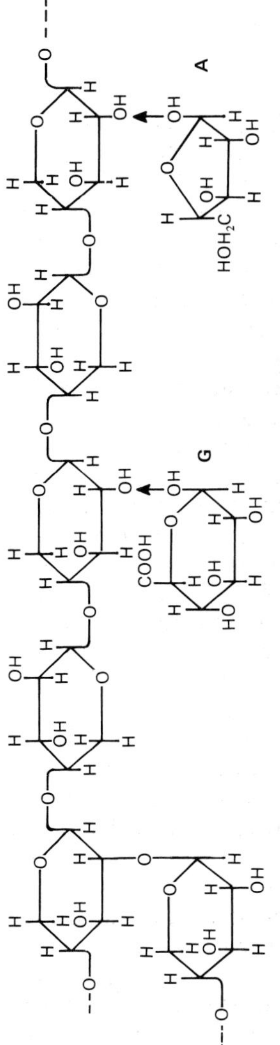

FIG. 2.18. Typical structural features of wheat xylan. The more soluble endosperm xylans are richer in L-arabinose whereas bran xylans are less soluble, more highly branched and are richer in D-glucuronic acid. Arrows mark the points of attachment of any L-arabinose (A) and D-glucuronic acid (G) residues.

15 per cent, lipid 5 per cent and inorganic and other substances making up the remainder. Much less hemicellulose (about 5 per cent) is found in the endosperm which is the basis of white flour. Nevertheless various effects of the hemicelluloses have been noted on the milling and baking characteristics of particular flours.

The nutritional benefits of high fibre diets are highly publicised in relation to many of the diseases which are characteristic of modern, urban man and his highly refined food. The effects on heart disease and related conditions that are made much of are probably derived from the physical interference of the bulky, indigestible fibre in digestion and absorption of fats from the small intestine. Presumably eating less fat would achieve the same result less wastefully. It is much clearer that the bulk that fibre gives to the contents of the large intestine is highly beneficial to bowel function. The reduced incidence of bowel cancer amongst those whose diet is rich in fibre is probably due to the effect of the increased bulk in reducing the time that potential carcinogens derived from our food spend in the bowel.

One should beware of automatically regarding a high dietary fibre level as beneficial. One component of bran, phytic acid, inositol hexaphosphate [2.37], which is the phosphate reserve for the germinating seed, and a potent complexer of divalent cations, has been implicated in calcium and zinc deficiency diseases amongst children who consume very little milk and large quantities of baked goods made from wholemeal flour.

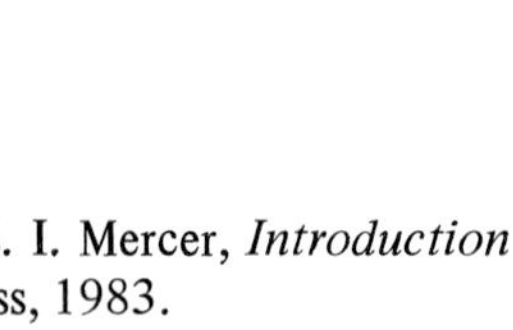

2.37

Further reading

T. W. Goodwin and E. I. Mercer, *Introduction to plant biochemistry*. Oxford: Pergamon Press, 1983.

Polysaccharides in food (J. M. V. Blanshard and J. R. Mitchell eds). London: Butterworths, 1979.

R. S. Shallenberger and G. G. Birch, *Sugar chemistry*. Westport: AVI Publishing Co, 1975.

D. A. Rees, *Polysaccharide shapes*, London: Chapman and Hall, 1977.

Wheat chemistry and technology (Y. Pomeranz ed), 2nd edn. St Paul, Minnesota: American Association of Cereal Chemists, 1971.

Starch and its derivatives (J. A. Radley ed), 4th edn. London: Chapman & Hall, 1968.

Water activity: influences on food quality (L. B. Rockland and G. F. Stewart eds). London: Academic Press, 1981.

Sugar: science and technology (G. G. Birch and K. J. Parker eds). London: Applied Science Publishers, 1979.

Carbohydrates, 2nd edn, Unilever Educational Booklet, 1975.

3. Lipids

The term 'lipid' refers to a group of substances that is even less clearly defined than the carbohydrates. It generally denotes a heterogeneous group of substances, associated with living systems, which have the common property of insolubility in water but solubility in non-polar solvents such as hydrocarbons or alcohols. Included in the group are the oils and fats* of our diet together with the so-called phospholipids associated with cell membranes. These substances have in common that they are esters of long chain fatty acids but there are many other lipids that lack this structural feature. They include the steroids and terpenes but with the exception of cholesterol (and its long chain fatty acid esters!) where these substances do have significance to food they will be found under the more relevant headings of vitamins, pigments or flavour compounds.

Fatty acids

Monocarboxylic, aliphatic, fatty acids are the structural components common to most of the lipids that interest food chemists and since many of the properties of food lipids can be accounted for directly in terms of their component fatty acids they will be considered in some detail. Almost without exception the fatty acids that occur in foodstuffs contain an even number of carbon atoms in an unbranched chain, *eg* lauric or dodecanoic acid, [3.1]. It is worthwhile to master both the

$$CH_3{-}CH_2{-}CH_2{-}CH_2{-}CH_2{-}CH_2{-}CH_2{-}CH_2{-}CH_2{-}CH_2{-}CH_2{-}COOH$$

3.1

trivial names, usually derived from an important source, as well as the systematic names of the principal fatty acids. Besides the saturated

*There is no formal distinction between oils and fats. The former are liquid, and the latter solid, at ambient temperatures.

fatty acids of which lauric acid (dodecanoic acid) is an example unsaturated fatty acids having one, two or sometimes up to six double bonds are common. The double bonds are almost invariably *cis* and when the fatty acid has two or more double bonds they are 'methylene interrupted':

$$—CH_2—CH{=}CH—CH_2—CH{=}CH—CH_2—$$

rather than conjugated:

$$—CH_2—CH{=}CH—CH{=}CH—CH_2—.$$

Thus α-linoleic acid* (systematically all-*cis*-9,12,15-octadecatrienoic acid) has the structure:

$$CH_3—CH_2—CH{=}CH—CH_2—CH{=}CH—CH_2—CH{=}CH—(CH_2)_7—COOH$$

The system used for the identification of double bond positions will be apparent by comparison of the structure with the systematic name. The structure of a fatty acid can be indicated by a convenient shorthand form in which the total number of carbon atoms is written followed by a colon and then the number of double bonds. The positions of the double bonds can be shown after the symbol △. Thus for example α-linoleic acid would be written simply '18:3△9,12,15'.

Table 3.1 gives the names and structures of many of the fatty acids commonly encountered in food lipids.

Table 3.1. The fatty acids commonly found in foodstuffs.

Systematic name	Common name	'Structure'
Saturated acids		
n-butanoic	butyric	4:0
n-hexanoic	caproic	6:0
n-octanoic	caprylic	8:0
n-decanoic	capric	10:0
n-dodecanoic	lauric	12:0
n-tetradecanoic	myristic	14:0
n-hexadecanoic	palmitic	16:0
n-octadecanoic	stearic	18:0
n-eicosanoic	arachidic	20:0
n-docosanoic	behenic	22:0
Unsaturated acids		
cis-9-hexadecenoic	palmitoleic	16:1△9
cis-9-octadecenoic	oleic	18:1△9
cis,cis-9,12-octadecadienoic	linoleic	18:2△9,12
all *cis*-9,12,15-octadecatrienoic	α-linolenic	18:3△9,12,15
all *cis*-5,8,11,14-eicosatetraenoic	arachidonic	20:4△5,8,11,14
all *cis*-7,10,13,16,19-docosapentaenoic	clupanodonic	22:5△7,10,13,16,19

*A rare isomer with double bonds at the 6, 9 and 12 positions is known as γ-linolenic acid.

The oils and fats are obviously the lipids that most interest the food chemist. These consist largely of mixtures of triglycerides, *ie* esters of the trihydric alcohol, glycerol (propane-1,2,3-triol) [3.2], and three fatty acid residues which may or may not be identical. 'Simple' triglyceride molecules have three identical fatty acid residues; 'mixed' triglycerides have more than one species of fatty acid. Thus a naturally occurring fat will be a mixture of quite a large number of mixed and simple triglycerides. It is important to remember that organisms achieve a desirable pattern of physical properties for the lipids of, say, their cell membranes or adipose tissue by utilizing an appropriate, and possibly unique, *mixture* of a number of different molecular species, rather than by utilizing a single molecular species which alone has the desired properties, as is the usual tactic with proteins or carbohydrates.

$$\begin{array}{ll} CH_2OH & 1 \\ | & \\ CHOH & 2 \\ | & \\ CH_2OH & 1' \text{ or } 3 \end{array}$$

3.2

The relative abundance and importance of the various fatty acids will be appreciated from an inspection of Table 3.2 where the molar proportions of the fatty acids in a number of fats and oils are presented. The origins of the names of some fatty acids will also be apparent.

The relationship between the fatty acid composition of a fat and its melting properties will be apparent in Table 3.2. The low melting temperature that characterises the oils is associated with either a high proportion of unsaturated fatty acids, *eg* corn oil and olive oil, or a high proportion of short chain fatty acids, *eg* milk fat and coconut oil. The MPs of the fatty acids are shown in Table 3.3.

The insertion of a *cis* double bond has a dramatic effect on the shape of the molecule, introducing a kink of about 42° into the otherwise straight hydrocarbon chain as shown in *Fig. 3.1*. The insertion of a *trans* double bond has very little effect on the conformation of the chain and therefore very little effect on the melting temperature. For example elaidic acid, the *trans* isomer of oleic acid, has similar physical properties to stearic acid.

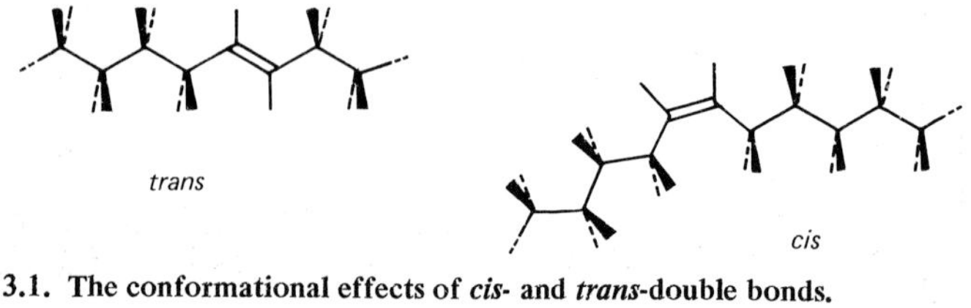

trans

cis

FIG. 3.1. The conformational effects of *cis*- and *trans*-double bonds.

Table 3.2. The fatty acid compositon of fats and oils.

Fatty acid	Beef fat	Lard	Cow's milk fat	Hard margarine	Soft margarine	Cod liver oil	Maize (corn) oil	Cocoa† butter	Coconut oil	Palm oil	Olive oil
4:0			9								
6:0			5								
8:0			2	6	4				12		
10:0			4						8		
12:0			3						49		
14:0	4	2	10	6	1	4			16	1	
14:1	1		2								
15:*	2		1			1					
16:0	28	26	23	26	13	13	14	29	7	48	11
16:1	5	4	2	5		18					1
17:*	1		1			2					
18:0	20	15	12	9	8	4	2	35	2	6	3
18:1	34	44	23	29	21	26	34	32	5	34	79
18:2	3	9	2	8	50	2	48	3	1	11	5
18:3	2		1	1	1	1	1				1
18:4						3					
20:0							1	1			
20:1				5	1	9					
20:4						1					
20:5						6					
22:1				5	1	5					
22:5						1					
22:6						14					

The figures here are typical values expressed in mole percent. There is often considerable variation in composition between samples of the same fat due to, for example, variations in the animal's diet or the conditions for plant growth. Trace components, *ie* those representing less than 0.5 per cent, have been ignored.

* The sums of straight chain, branched, saturated and unsaturated acids.

† The major component of chocolate.

Table 3.3. Fatty acid melting points.

Fatty acid	MP °C	Fatty acid	MP °C
Butyric	−7.9	Elaidic	43.7
Caproic	−3.4		
Caprylic	16.7	Oleic	10.5
Capric	31.6	Linoleic	−5.0
Lauric	44.2	Linolenic	−11.0
Myristic	54.1		
Palmitic	62.7	Arachidonic	−49.5
Stearic	69.6		
Arachidic	75.4		

The milk fats of ruminants are characterised by their high proportion of short chain fatty acids. These are derived from the anaerobic fermentation of carbohydrates such as cellulose by the micro-organisms of the rumen. These microorganisms are also the source of the very small proportions of branched chain fatty acids that occur in cow's milk fat. Branched chain fatty acids are usually of *iso* series which have their hydrocarbon chain terminated:

$$\begin{matrix} CH_3 \searrow \\ \quad CH{-}CH_2{-}{-}{-} \\ CH_3 \nearrow \end{matrix}$$

or the *anteiso* series:

$$\begin{matrix} CH_3{-}CH_2{-}\underset{\displaystyle CH_3}{\underset{|}{CH}}{-}CH_2{-}{-}{-} \end{matrix}$$

Human milk fat, like that of other non-ruminant species such as the pig, is rich in linoleic acid. Although no direct benefits of this to human babies have yet been identified it will be surprising if there are none in view of the importance of linoleic acid in human nutrition (see later in this chapter).

Cod liver oil is most familiar to us as a source of vitamin D but fish oils are particularly interesting for the diversity of long chain, highly unsaturated, fatty acids they contain. The extremely cold environments of fish such as cod and herring are perhaps one reason for the nature of fish lipids.

There are a number of fatty acids with unusual structures which are characteristic of particular groups or species of plants. For example petroselenic acid (18:1△11 *cis*) is found in the seed oils of celery, parsley and carrots. Ricinoleic acid [3.3] makes up about 90 per cent of the fatty acids of castor oil.

$$CH_3(CH_2)_5\underset{\displaystyle OH}{\underset{|}{C}}HCH_2CH{=}CH(CH_2)_7COOH$$

3.3

Fatty acids containing cyclopropene rings are mostly associated with bacteria but traces of sterculic acid [3.4] and malvalic acid [3.5] are found in cottonseed oil. They are toxic to non-ruminants and the residual oil in cottonseed meal is sufficient to have an adverse effect on poultry fed on the meal. It must be assumed that the low levels man has consumed for many years in salad dressings and margarine manufactured from cottonseed oil have had no deleterious effects.

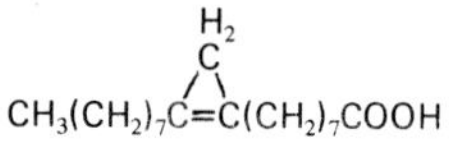

$$CH_3(CH_2)_7\overset{\overset{H_2}{C}}{C{=}C}(CH_2)_7COOH$$

3.4

$$CH_3(CH_2)_7\overset{\overset{H_2}{C}}{C{=}C}(CH_2)_6COOH$$

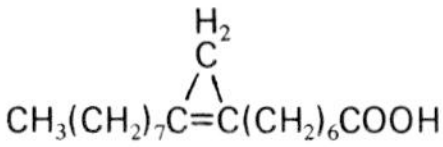

3.5

Erucic acid, *cis*-13-docosenoic acid, is a characteristic component of the oils extracted from the seeds of members of the genus *Brassica*, notably varieties of *B. napus* (in Europe) and *B. campestris* (in Canada and Pakistan) which are known as rape. Rapeseed meal is an important animal feeding stuff and rapeseed oil is important for both human consumption and industrial purposes such as lubrication. Most rapeseed oil used to contain some 20–25 per cent erucic acid, to which was attributed a number of disorders concerning the lipid metabolism of rats fed high levels of the oil. The problem of possible toxicity to humans has been largely eliminated by plant breeders who have developed strains of rape giving oil with acceptably low erucic acid levels (less than 5 per cent) which are utilized in margarine manufacture. Oils rich in erucic acid remain important for many industrial applications.

The fatty acids, in the form of the triglycerides of the dietary fats and oils, provide a major proportion of our energy requirements as well as, when in excess, contributing to the unwelcome burden of superfluous adipose tissue that so many of us carry. In recent years we have begun to appreciate that certain dietary fatty acids have a more particular function in human nutrition. Rats fed on a totally fat free diet show a wide range of acute symptoms affecting the skin, vascular system, reproductive organs and lipid metabolism. Although no corresponding disease state has ever been recorded in a human patient similar skin disorders have occurred in children subjected to a fat free diet. The symptoms in rats could be eliminated by feeding linoleic or arachidonic acids (which for a time became known as vitamin F in consequence) and it is generally accepted that 2–10 g of linoleic acid per day will meet an adult human's requirements.

The identification of these two 'essential fatty acids' in the 1930s preceded by some 25 years their identification as precursors of a group of animal hormones, the *prostaglandins*. Although animal tissues are unable to synthesise either of these two fatty acids they readily convert the C_{18} acid to the C_{20} acid as shown in *Fig. 3.2*.

The many different prostaglandins all have structures closely related to the example E_2 shown in *Fig. 3.2*. The reasons for the stringent requirements for the positions of the double bonds in essential fatty acids are clearly demonstrated in the simplified pathway presented in *Fig. 3.2*.

Other prostaglandins vary in the degree of reduction of the ring oxygens and presence of double bonds in the chain. Details of their

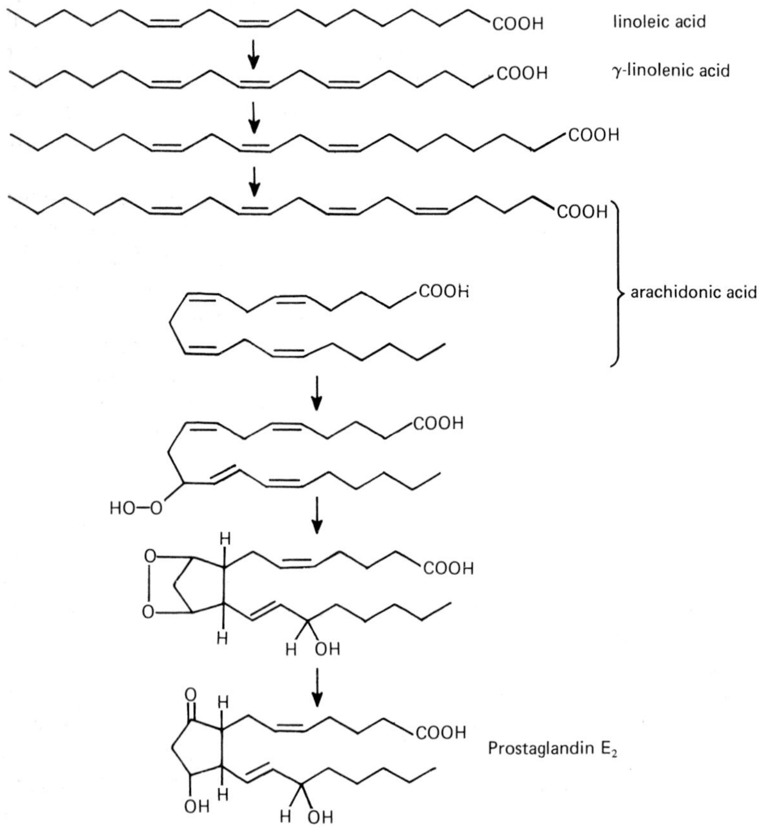

FIG. 3.2. The biosynthesis of prostaglandin E_2 from linoleic acid.

numerous physiological activities are still accumulating in the scientific literature but they are best known for their involvement in inflammation and the contraction of smooth muscle. Aspirin is an inhibitor of the first oxygenation step in prostaglandin biosynthesis.

The involvement of dietary fatty acids in the occurrence of *atherosclerosis* is a complex issue. The artery wall lesions that characterise the disease are undoubtedly associated with elevated levels in the blood of cholesterol and abnormal levels of the lipoproteins that carry cholesterol. There is also little doubt that in many individuals the blood cholesterol level is markedly increased following consumption of fats that are rich in saturated fatty acids. The apparent value of polyunsaturated fatty acids in combating arterial disease has been made much of in recent years; particularly by the vegetable oil industry and to the embarrassment of the dairy industry. It seems, however, that what is required is a reduction of the intake of saturated fatty acids rather than an increased intake of polyunsaturated ones whose intrinsically beneficial role is probably confined to their 'essential fatty acid' function. It should not be overlooked that plant fats and oils are neither necessarily nor exclusively rich in polyunsaturated fatty acids, especially after they have been hydrogenated (see below) for margarine manufacture. Rather curiously, dietary cholesterol seems to be quite without influence on blood cholesterol levels.

The mechanisms underlying the exceedingly complex interactions between diet, stress, smoking, genetic and other factors in the incidence of heart disease are far from fully understood and the fact that the commercial interests of two competing sectors of the food industry are at stake may well hinder rather than help their elucidation.

Unsaturated fatty acids take part in a number of chemical reactions that are important to the food scientist, the hydrocarbon chain of the saturated fatty acids being essentially chemically inert under the conditions encountered in food. These reactions do not involve the carboxyl group and since most occur whether or not the fatty acid is esterified with glycerol it is convenient to deal with them at this point rather than when the particular properties of the fats and oils are considered.

Halogens react readily with the double bonds of fatty acids and the stoichiometry of the reaction has been utilized for many years as a guide to the proportion of unsaturated fatty acids in a fat. In many laboratories the *iodine number*, *ie* the number of grams of iodine that react with 100 g of fat, is still routinely determined, usually by one of the standardised procedures such as Wij's method, as an aid in the identification of unknown fats and in the monitoring of processes such as hydrogenation. In modern laboratories complete data on the fatty acid composition is readily obtained by the gas chromatography of the volatile methyl esters of the fatty acids obtained by saponification and methylation of the fat.

In the presence of hydrogen and a suitable catalyst the double bonds of unsaturated fatty acids are readily hydrogenated, a reaction at the heart of modern* margarine manufacture. The aim of hydrogenation is to convert a liquid vegetable or fish oil to a fat with a butter-like consistency by reducing the degree of unsaturation of its component fatty acids.

After preliminary purification to remove polar lipids and other substances which tend to 'poison' the catalyst, the oils are exposed to hydrogen gas at high pressures and temperatures (2–10 atmos., 160–220 °C) in the presence of 0.01–0.2 per cent finely divided nickel. Under these conditions the reduction of the double bonds is fairly selective. Those furthest from the ester link of the triglycerides and those belonging to the most highly unsaturated fatty acid residues are most reactive. This fortunate selectivity thus results in trienoic acids being converted to dienoic and dienoic to enoic acids rather than an accumulation of fully saturated acids. Of course the margarine manufacturer regulates the degree of hydrogenation to give the particularly desired characteristics to the finished product.

The hydrogenation reactions are by no means as straightforward as one might assume. *Cis–trans* isomerisation and double bond migration occur at the high temperatures involved and together with the occasional hydrogenation of the 'wrong' bonds give rise to significant proportions of unnatural fatty acids such as 18:1△12 in margarine. The presence of *trans* unsaturated fatty acids is highly characteristic of hydrogenated oils and may be detected by their characteristic infrared absorption bands (965–990 cm^{-1}). A typical margarine may well contain up to 20 per cent of fatty acids with *trans* double bonds. Of these elaidic acid (18:1 *trans*△9) is the best known.

The metabolism of *trans* fatty acids has received very little attention from biochemists but in spite of being part of our diet since the early years of this century no deleterious effects have been ascribed to them. The usual metabolic route for the breakdown of fatty acids, mitochondrial β-oxidation, includes at one stage an intermediate which has a *trans* double bond so it seems likely that we are able to take these unnatural fatty acids in our metabolic stride.

Rancidity is a familiar indication of the deterioration of fats and oils. In dairy fats rancidity is usually the result of hydrolysis of the triglycerides (lipolysis) by micro-organisms so that odorous short chain fatty acids are liberated. In other fats and oils, and the fatty parts of meat and fish rancidity is the result of autoxidation of the unsaturated fatty acids. The course of the reaction is essentially similar in

---

*The first margarine was invented by a French chemist Miège Mouriès in 1869 as a butter substitute. It was a mixture of milk, chopped cow's udder tissue and a low melting fraction of beef fat!

free or esterified fatty acids and occurs in three stages. The initiation reactions give small numbers of highly reactive molecules that have unpaired electrons, free radicals. In the propagation reactions, oxygen reacts with these to give fatty acid hydroperoxides which then break down to generate more free radicals which maintain the chain of reactions. When the free radical species reach sufficiently high concentrations they tend to react together in termination reactions to give the stable end products which are characteristic of rancid fat. A similar sequence of reactions occurs in many vegetable tissues, catalysed by the enzyme lipoxygenase, which give rise to off-flavours. The reactions of the autoxidation sequence are outlined in *Fig. 3.3* and certainly merit detailed consideration.

Until very recently there was great uncertainty about the nature of the initiation reactions. However it is now widely accepted that it is not oxygen in its ordinary ground state, so-called triplet oxygen $^3O_2$, that reacts with the double bond but the high energy, short lived and highly reactive singlet oxygen $^1O_2$. Singlet oxygen molecules arise from triplet oxygen in a number of reactions but most relevant to food systems are those in which $^3O_2$ reacts with pigments such as chlorophyll, riboflavin (very important in fresh milk deterioration) and haem in the presence

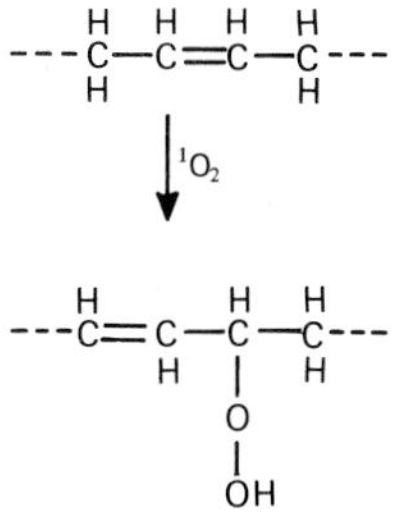

```
   H  H  H  H
---C—C=C—C---
   H        H
      |
      | 1O2
      ↓
   H     H  H
---C=C—C—C---
      H  |  H
         O
         |
         OH
```

FIG. 3.3(i). The autoxidation fatty acids - the initiation stage. $^1O_2$ reacts directly with a fatty acid double bond to give a hydroperoxide and the double bond migrates and takes up the *trans* configuration.

of light. Although the proportion of hydroperoxide initially formed in a fat is exceedingly small, the subsequent breakdown of the hydroperoxide yields free radical species that are able to propagate a chain reaction. These free radicals are then able to abstract H$^{\bullet}$, *ie* a hydrogen radical from the labile α-methylene groups of monoenoic and poly-

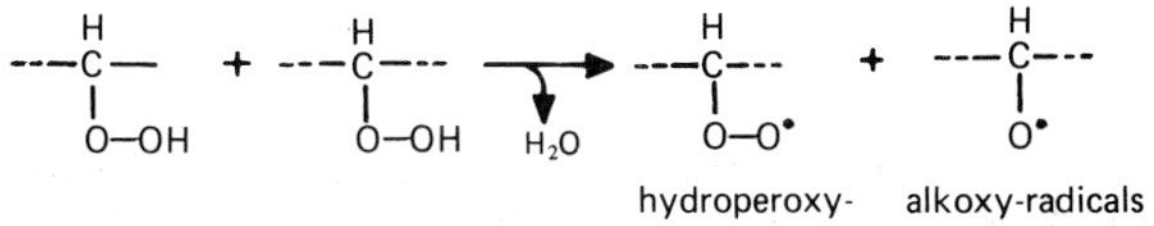

```
   H            H                 H              H
---C—    +  ---C---   ——→     ---C---    +   ---C---
   |            |       ↓         |              |
   O—OH         O—OH   H2O        O—O•           O•
                              hydroperoxy-   alkoxy-radicals
```

FIG. 3.3(ii). The autoxidation of fatty acids - initial hydroperoxide breakdown.

With a monoenoic acid such as oleic acid:

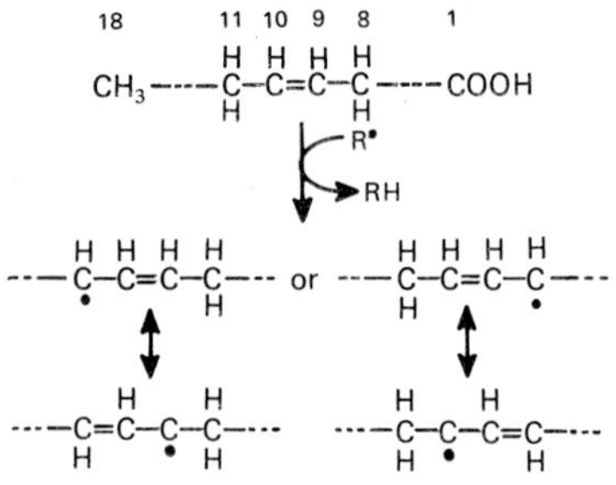

```
 18        11 10  9  8        1
           H  H  H  H
CH3----- C--C==C--C ----COOH
           H        H
                 | R•
                 |--> RH
                 v
   H  H  H  H                H  H  H  H
---C--C==C--C---    or   ---C--C==C--C---
   •        H                H        •
       ↕                         ↕
      H     H                H     H
---C==C--C--C---        ---C--C--C==C---
   H     •  H                H  •     H
```

With a dienoic acid such as linoleic acid:

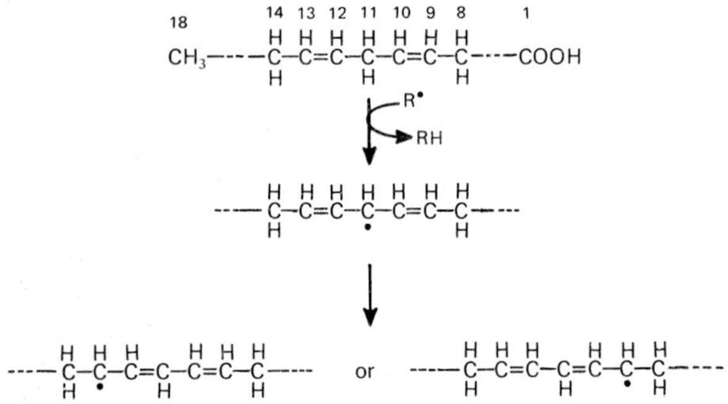

```
 18         14 13 12 11 10  9  8         1
            H  H  H  H  H  H  H
CH3-----  C--C==C--C--C==C--C ----COOH
            H        H        H
                     | R•
                     |--> RH
                     v
            H  H  H  H  H  H  H
     ----  C--C==C--C--C==C--C ----
            H        •        H
                     |
                     v
   H  H  H     H  H  H                  H  H  H     H  H  H
---C--C--C==C--C==C--C---     or    ---C--C==C--C==C--C--C---
   H  •        H     H                 H        H     •  H
```

In each case a hydroperoxide results by reaction with triplet oxygen and then another fatty acid residue:

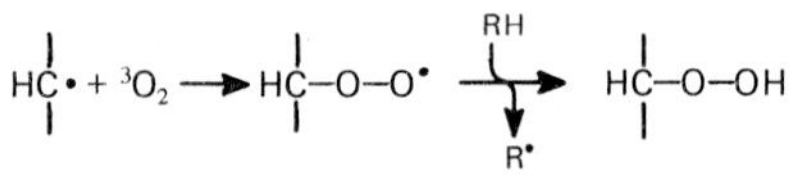

$$\mathrm{HC}\cdot + {}^{3}O_2 \longrightarrow \mathrm{HC{-}O{-}O}\cdot \xrightarrow[\ R\cdot\ ]{RH} \mathrm{HC{-}O{-}OH}$$

FIG. 3.3(iii). The autoxidation of fatty acids – the formation of hydroperoxides in the propagation phase.

---

enoic fatty acid residues. As shown in *Fig. 3.3(iii)* a great diversity of hydroperoxides ensues from the isomerisations that occur and the fat absorbs considerable quantities of oxygen from the atmosphere. Hydroperoxides accumulate in the autoxidising fat but as their concentration builds up their breakdown products become increasingly important. The alkoxy radicals that arise as shown in *Fig. 3.3(ii)* give rise to aldehydes, ketones and alcohols and a continuing supply of free

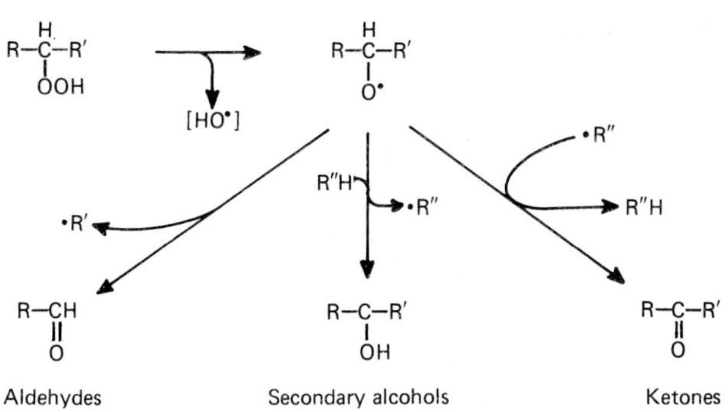

FIG. 3.3(iv). The autoxidation of fatty acids – hydroperoxide breakdown.

radicals to maintain the chain of reactions. Transition metal cations are important catalysts of hydroperoxide breakdown:

$$\begin{aligned} R{-}OOH + M^{+} &\longrightarrow R{-}O^{\cdot} + OH^{-} + M^{2+} \\ R{-}OOH + M^{2+} &\longrightarrow R{-}OO^{\cdot} + H^{+} + M^{+} \\ \hline 2R{-}OOH &\longrightarrow R{-}O^{\cdot} + R{-}OO^{\cdot} + H_2O \end{aligned}$$

Aldehydes arising from cleavage of the carbon chain on either side of the alkoxy radical are the source of rancid fat's characteristic odour. For example, the 9-hydroperoxide from linoleate will give either 2-nonenal or 2,4-decadienal as shown in *Fig. 3.3(v)*. One particular compound, malonaldehyde [3.6], results from cleavage at both ends of a diene system. It forms a pink colour with thiobarbituric acid which forms the basis of a useful method for assessing the deterioration of fats. Other tests include measurements of the carbonyl content using 2,4-dinitrophenylhydrazine, the hydroperoxide content by reaction with iodine and the characteristic absorbances of conjugated dienes and trienes at 230 nm and 270 nm respectively.

$$\underset{\substack{\Vert \\ O}}{HC}{-}CH_2{-}\underset{\substack{\Vert \\ O}}{CH}$$

3.6

The use of tests such as these, as well as monitoring the oxygen uptake, has shown that the course of oxidation of a fat is marked by an induction period of slow oxygen uptake (while the level of initiating free radicals builds up) followed by a period of rapid oxidation. Usually a pronounced rancid odour is not detectable until the rapid oxidation phase is well established. Some oils do develop off-flavours and odours even when very little oxidation has occurred. Soya bean oil is particularly susceptible to this phenomenon, termed

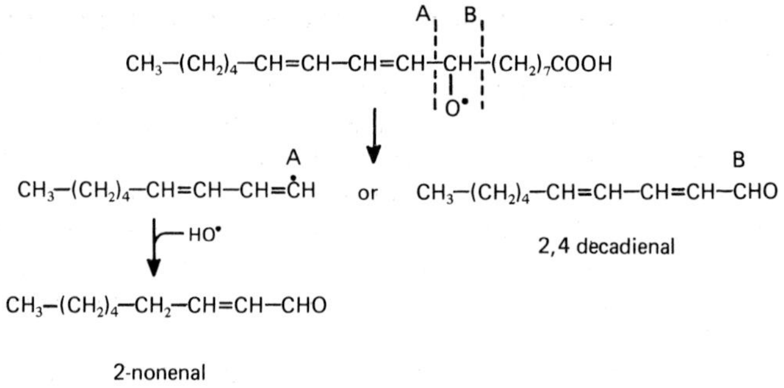

$$CH_3{-}(CH_2)_4{-}CH{=}CH{-}CH{=}CH\overset{A}{\vdots}\underset{|}{CH}\overset{B}{\vdots}(CH_2)_7COOH$$
$$\underset{O^\bullet}{}$$

$$\downarrow$$

A: $CH_3{-}(CH_2)_4{-}CH{=}CH{-}CH{=}\dot{C}H$ or B: $CH_3{-}(CH_2)_4{-}CH{=}CH{-}CH{=}CH{-}CHO$

2,4 decadienal

$$\downarrow HO^\bullet$$

$CH_3{-}(CH_2)_4{-}CH_2{-}CH{=}CH{-}CHO$

2-nonenal

FIG. 3.3(v). The autoxidation of fatty acids – formation of unsaturated aldehydes.

reversion, apparently due to its content of highly labile linoleic acid. Amongst the compounds that have been implicated in the 'flavour' of reverted soya bean oil are *cis*-3-hexenal, diacetyl (butanedione), 2,3-pentadione and 2,4-pentadienal. Reduction in the linolenic acid content of seed oils is an important objective for plant breeders.

Fats and oils exposed to the atmosphere and heating over a long period show the final stage of the oxidation sequence, polymerisation. The highly unsaturated oils used in paint show the phenomenon even more readily! The cross-linking reactions can be of various types. Free radicals may react directly together:

$$R^\bullet + R^\bullet \rightarrow R{-}R$$
$$R^\bullet + ROO^\bullet \rightarrow R{-}O{-}O{-}R$$
$$ROO^\bullet + ROO^\bullet \rightarrow R{-}O{-}O{-}R + O_2$$

or with other alkenic systems:

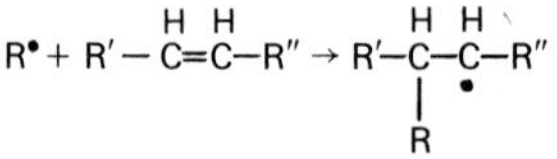

$$R^\bullet + R'{-}\overset{H}{C}{=}\overset{H}{C}{-}R'' \rightarrow R'{-}\underset{\underset{R}{|}}{\overset{H}{C}}{-}\overset{H}{\dot{C}}{-}R''$$

or Diels–Alder reactions may occur to give cyclic structures:

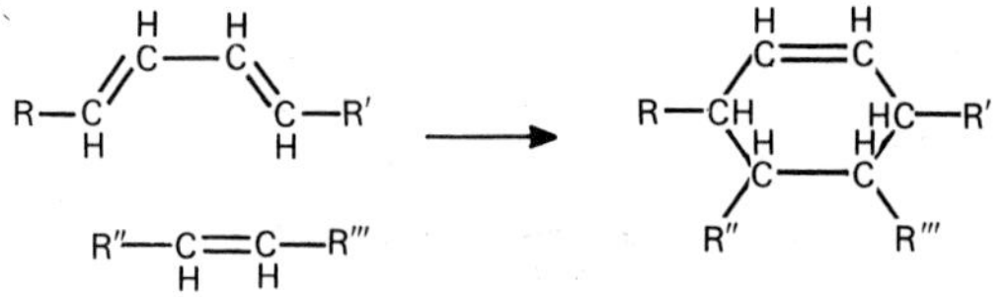

$$R{-}\underset{H}{C}{=}\overset{H}{C}{-}\overset{H}{C}{=}\underset{H}{C}{-}R' \quad + \quad R''{-}\underset{H}{C}{=}\underset{H}{C}{-}R''' \longrightarrow \text{cyclohexene: } R{-}CH,\ \overset{H}{C}{=}\overset{H}{C},\ HC{-}R',\ \overset{H}{C}(R''){-}\overset{H}{C}(R''')$$

Polymerization reactions are particularly associated with frying oils where prolonged use leads to high molecular weight compounds which cause foaming and increased viscosity. Discarded oils are sometimes found to contain as much as 25 per cent of polymerized material. The chain reaction character of the autoxidation reactions provides the explanation for why a batch of oil approaching the end of its useful life should be discarded in its entirety rather than diluted with fresh oil.

The question of the toxicity of deteriorated frying oils has not been satisfactorily resolved. Experiments with laboratory animals fed on autoxidised fats do demonstrate adverse effects but many doubts have been expressed as to the relevance of these results to human nutrition. There is very little evidence to suggest that oxidised fats cause cancer in humans. Perhaps the adverse effects of a diet over-rich in fats generally swamp these more subtle influences.

Although the development of rancidity in bulk fats and oils can be retarded by careful processing procedures avoiding high temperatures, metal contamination *etc* such measures are never wholly adequate. Fatty foods such as biscuits and pastry are particularly susceptible to rancidity as their structure necessarily exposes the maximum surface of the fat to the atmosphere. The shelf life of these types of foods (on the housewife's shelf as well as the shopkeeper's) is massively extended by the inclusion of antioxidants in many of the fats we buy, particularly lard. The structures of some important antioxidants are shown in *Fig. 3.4*. They appear to work by blocking the propagation process, the

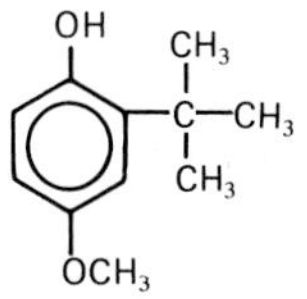

Butylated hydroxyanisole
(2-tert-butyl-4-methoxyphenol)
(BHA)

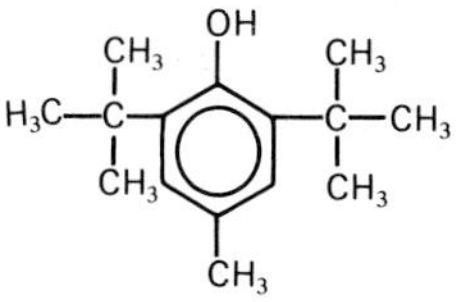

Butylated hydroxytoluene
(2,6-bis(1,1-dimethylethyl)-4-methylphenol)
(BHT)

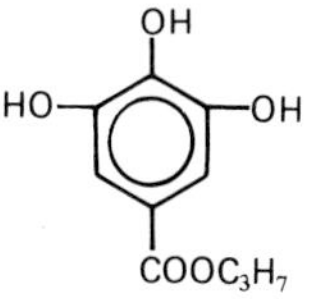

Propyl gallate
(3,4,5-trihydroxy-benzoic acid propyl ester) (PG)

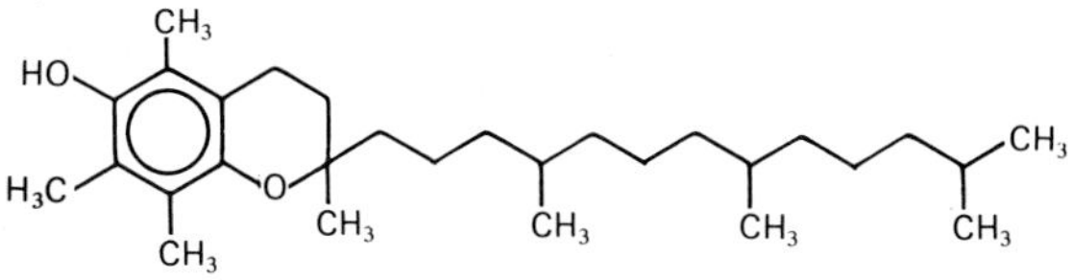

α-Tocopherol
(Vitamin E)

FIG. 3.4. Some antioxidants important in food lipids.

antioxidant (shown here as AH) donating a hydrogen atom to a free radical such as $ROO^{\bullet}$:

$$AH + ROO^{\bullet} \rightarrow ROOH + A^{\bullet}$$

The antioxidant free radicals are stabilised by resonance:

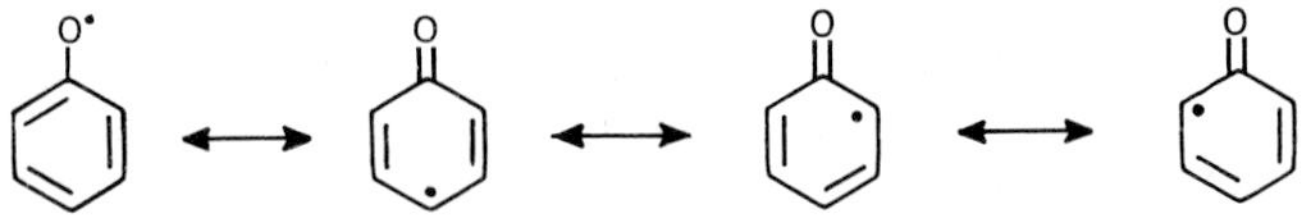

and are inactive in the continuation of the reaction chain. They appear to enter termination reactions such as

$$A^{\bullet} + A^{\bullet} \longrightarrow A{-}A$$

or

$$A^{\bullet} + ROO^{\bullet} \longrightarrow ROO{-}A$$

Antioxidants do not reduce the ultimate degree of rancidity, rather they lengthen the induction period in rough proportion to their concentration. For example, in an experiment, pastry made with lard lacking antioxidant smelt rancid after 7 days in air at room temperature whereas 44 days elapsed before a similar smell developed in pastry made with lard containing 0.1 per cent butylated hydroxyanisole (BHA). The three synthetic antioxidants shown in *Fig. 3.4* (BHA, BHT and PG) are normally added to fats at rather lower levels (up to about 200 ppm). The tocopherols are antioxidants that occur in most plant tissues, as much as 0.1 per cent in vegetable oils. Animals require tocopherols (as vitamin E) in their diet and in their absence membrane lipids show effects that can be identified with autoxidation.

Triglycerides

Having considered the properties of the fats and oils from the point of view of the chemistry of their component fatty acids we can now examine them in terms of their component triglycerides. The first descriptions of the glyceride structure of fats assumed that all their component triglycerides were simple. Thus a fat containing palmitic (hexadecanoic), stearic (octadecanoic) and oleic (*cis*-octadec-9-enoic) acids would be a mixture of three triglyceride species tripalmitin, tristearin and triolein. The first attempts to separate the component glycerides of fats, by the laborious process of fractional crystallisation from acetone (propanone) solutions at low temperatures, made it clear that much greater numbers of species of triglycerides occurred than would be expected from this simple concept. Fats and oils became recognised as clearly defined mixtures of mixed and simple triglycerides.

If we reconsider our simple fat with only three fatty acids then there should be 27 possible triglycerides. However, as the 1 and 1′ positions of the glycerol molecule are usually treated as indistinguishable the total comes down to 18. The range of possibilities is illustrated by

Table 3.4 whose data have been reduced to manageable proportions by grouping the fatty acids.

The most striking feature of Table 3.4 is that although these two fats have rather similar fatty acid compositions their triglyceride compositions are very different. In lard there is a definite tendency for unsaturated fatty acids to occupy the outer positions of the glycerol molecule whereas in cocoa butter the reverse is true. This difference is in fact a general one between animal and plant fats as is borne out by the further triglyceride compositions given in Table 3.5. If the fatty acid residues at the two outer positions are different the triglyceride molecule is asymmetric and optical activity would be expected.

It has been possible to show that a particular species of triglyceride in a natural fat does occur as a single enantiomer which implies that

Table 3.4. Typical triglyceride compositions of lard and cocoa butter.

Position on the glycerol molecule			Mole per cent	
1	2	1	*Lard*	*Cocoa butter*
Fully saturated				
P	P	P	0.3	0.3
P	P	S	1.7	0.9
S	P	S	2.4	0.6
P	S	P	0.2	0.2
P	S	S	0.1	0.4
S	S	S	0.2	0.3
One unsaturated residue				
P	P	U	6.5	0.1
S	P	U	18.9	0.2
P	S	U	0.4	0.1
S	S	U	1.2	0.1
P	U	P	0.1	14.1
P	U	S	0.7	39.3
S	U	S	1.1	27.4
Two unsaturated residues				
U	P	U	36.5	0
U	S	U	2.4	0
P	U	U	2.9	6.4
S	U	U	8.3	8.9
Fully unsaturated				
U	U	U	16.1	0.7

P – all saturated fatty acids with 16 carbon atoms or less.
S – all saturated fatty acids with 18 carbon atoms or more.
U – all unsaturated fatty acids.

Data obtained by the pancreatic lipolysis techniques by M. H. Coleman, 1961, *J. Am. Oil Chem. Soc.*, 1961, **38**, 685.

Table 3.5. Simplified triglyceride compositions (mole percent) of some animal and plant fats

Tryglyceride	Beef fat	Chicken fat	Palm oil	Groundnut oil
UUU	3	24	10	47
USU	2	7	2	1
SUU	17	38	36	41
SUS	32	16	34	9
SSU	16	10	9	1
SSS	30	5	8	1

Saturated fatty acids are collectively listed as S, unsaturated as U.

during triglyceride biosynthesis the 1 and 1′ positions are distinguished. In spite of this distinction the few studies that have been undertaken have shown only small differences in the 'compositions' of positions 1 and 1′. Milk fats are exceptional in that their triglycerides fall into three broad classes. In the first, all three positions are occupied by long chain fatty acids. The second class has long chain fatty acids at the 1 and 2 positions but short chain at 1′. The third class has medium chain fatty acids at 1 and 2 and medium or short chain fatty acids at the 1′ position.

The determination of the triglyceride composition of a fat is particularly difficult. The need to distinguish between isomeric pairs of glycerides such as 'SUS' and 'SSU', crucial to the difference between cocoa butter and lard for example, makes special demands on chromatographic procedures. Thin layer chromatography on silica gel impregnated with silver nitrate has been very successful as it will resolve these isomeric pairs. The silver ions are complexed by the double bonds of unsaturated triglycerides whose mobility is thus reduced.

The identity of a purified triglyceride can be established by means of pancreatic lipase. This enzyme specifically catalyses the hydrolysis of 1 and 1′ ester links of a triglyceride so that in the laboratory, as in the small intestine, the products from a triglyceride molecule will be the two fatty acids from the outer positions and one 2-monoglyceride:

$$\begin{array}{l} CH_2OC(=O){-}R' \\ | \\ CHOC(=O){-}R'' \\ | \\ CH_2OC(=O){-}R''' \end{array} + 2H_2O \longrightarrow \begin{array}{l} CH_2OH \\ | \\ CHOC(=O){-}R'' \\ | \\ CH_2OH \end{array} + R'COOH + R'''COOH$$

The application of the method to an entire fat depends on two assumptions (which other work shows to be largely valid):

(*i*) There is no distinction between the 1 and 1′ positions.
(*ii*) The nature of the fatty acid at one position of a particular triglyceride has no relationship to those at other positions, *ie* the fatty acids are randomly distributed with respect to each other and each position therefore has its own particular fatty acid composition.

Using these assumptions one may calculate the triglyceride composition from compositions of the 2 position (obtained from the monoglyceride fraction of the pancreatic lipolysis products) and of 1 and 1′ positions (obtained by subtraction of the 2 positions contribution to the composition of the entire fat).

Over the years many unsuccessful attempts have been made to devise formulae which would enable the prediction of the triglyceride composition of a fat from its fatty acid compositions. At the present time it is generally accepted that the '1,1′-random, 2-random' distribution theory, which is essentially the assumptions given above in the context of the pancreatic lipolysis method, is about as far as one can go towards a generally applicable prediction.

For the food chemist the melting and crystallisation characteristics of a fat are physical properties of prime importance. Although the melting points of pure triglycerides are a function of the chain lengths and unsaturation of the component fatty acids, much as one might expect, the melting behaviour of fats is rather complex. Since natural fats are mixtures, each component having its own melting point, a fat does not have a discreet melting point but rather a 'melting range'. At temperatures below this range all the component triglycerides will be below their individual melting points and the fat will be completely solid. At the bottom of the range the lowest melting types, those of lowest molecular weight or most unsaturated, will liquefy. Some of the remaining solid triglycerides will probably dissolve in this liquid fraction. As the temperature is raised the proportion of liquid to solid rises and the fat becomes increasingly plastic until, at the temperature arbitrarily defined as the melting point, there is no solid fat left.

A second complication is that triglycerides are polymorphic, *ie* they can exist in several different crystalline arrangements, each with its characteristic melting point, x-ray crystallographic pattern and infrared spectrum. The three principal forms are known as α, β and β'. The melting points of these forms of a number of triglyceride species are given in Table 3.6.

As an example of the formation of these different types we can consider tristearin. The α form can be obtained by rapid solidification of liquid tristearin; the β' form results from slow heating of the α form

Table 3.6. The melting points of the polymorphic forms of triglycerides (°C).

	α	β	β'
Tricaprin	–	–	32
Trilaurin	14	34	44
Trimyristin	32	44	56
Tripalmitin	44	56	66
Tristearin	54	64	73
Triolein	−32	−13	4

which melts and then solidifies at 64 °C; the β form can be obtained by recrystallisation from solvent. Although each form has been fairly well characterised in terms of its X-ray diffraction pattern and infrared spectrum very little is established about the actual arrangement of the triglyceride molecules in the crystals. Some features of the different crystalline arrangements are shown in *Fig. 3.5*. As far as is known in all three forms the 1 and 1′ fatty acids lie in the opposite direction to the 2 fatty acid. The α forms are the least well characterised. Their crystals are fragile platelets about 5 μm in size. The β' forms have smaller needle-like crystals while β forms have large crystals (up to 100 μm) which clump together to give coarse textures. *Figure 3.5* shows how subdivisions of these broad classes arise through different tilt angles and different patterns of overlap. The preferred crystal structure for a particular triglyceride is not easily predicted, being dependent on the chain lengths of the three fatty acid residues and the disruptive presence of double bonds. In a natural fat the occurrence of large numbers of different species of triglyceride will complicate the matter further but one polymorphic type usually predominates. For example in coconut oil, corn oil and lard β predominates whereas β' is the predominant type in palm oil, milk fats and herring oil.

The usefulness of a fat to a particular food application is crucially dependent on its melting and crystallisation characteristics. For example, fats to be spread on bread or blended with flour *etc* in cake or pastry mixes require the plasticity given by a wide melting range. The nature of the crystalline form also influences the creaming or shortening capability, *ie* the ability to incorporate air bubbles on mixing with flour or sugar. For this there should be an ample liquid fat matrix containing large numbers of very small crystals. Since small crystals are usually associated with β' forms the basis of the different applications of lard and butter in pastry, cakes and other baked goods now becomes clearer.

Cocoa butter is a good example of the importance of melting characteristics to a foodstuff – in this case chocolate. The special characteristic of chocolate is its sharp melting point, it 'melts in your mouth but

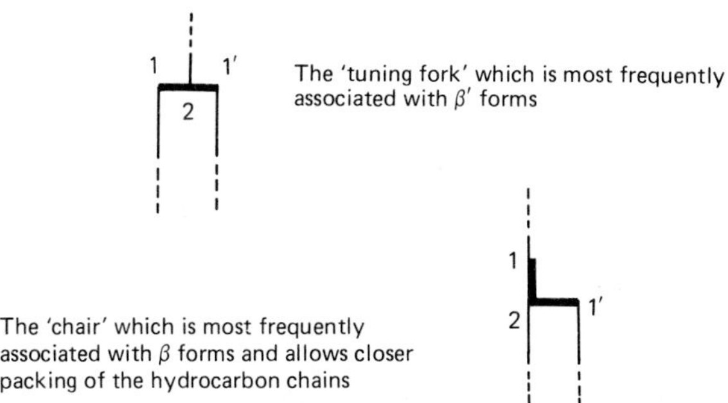

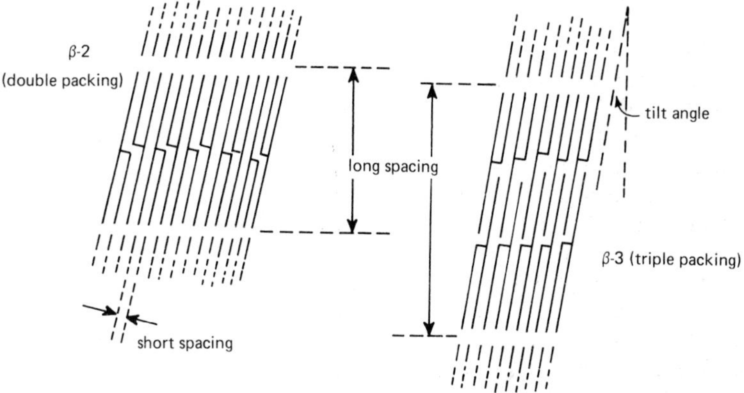

The dimensions of the long and short spacings and the size of tilt angle characterise the different types and are established by X-ray crystallography.

FIG. 3.5. Structural variations in the molecular arrangements of triglyceride polymorphic forms.

not in your hand!' Table 3.4 showed that 80 per cent of its triglycerides are of one type, SOS*, and the similarities of the melting points of the various triglycerides of this type are sufficient to ensure that cocoa butter has a narrow melting range. We also expect chocolate to have a very smooth texture and a glossy surface. Cocoa butter can occur in six different polymorphic states with melting points ranging from 17.3 °C to 36.4 °C. Only the fifth of these (a β-3 type, MP 33.8 °C) has the desired properties and the special skill of the chocolate maker

* S = Saturated, palmitic or stearic acids. O = Oleic acid.

lies in ensuring that the fat is in this particular state in the finished product. This is achieved by tempering. The liquid chocolate is cooled to initiate crystallisation and reheated to just below the melting point of the desired polymorphic type so as to melt out any of the undesirable types. The chocolate is then stirred at this temperature for some time so as to obtain a high proportion of the fat as very small crystals of the desired type when it is finally solidified in the mould or when it is coating biscuits or confectionery.

Chocolate that has been incorrectly tempered or subjected to repeated fluctuations in temperature, as in, for example, a shop window, develops bloom. This is a grey film which resembles mould growth but is actually caused by the transition of some of the fat to a more stable polymorphic form which crystallises out on the surface. The migration of triglycerides from the nut centres of chocolates or the crumb of chocolate coated biscuits can cause similar problems. Milk fat is an effective bloom inhibitor and is often included in small amounts in plain chocolates, besides accounting for about one quarter of the total fat in a typical milk chocolate. Cocoa butter is obviously a very difficult fat to handle successfully in a domestic kitchen and chocolate substitutes are marketed for home cake decorating *etc*. These contain fats such as hardened (*ie* partially hydrogenated) palm kernel oil. Although they have an inferior, somewhat greasy, texture these artificially modified fats only have a single polymorphic state and therefore do not present the housewife with the problems of tempering.

Polar lipids

The membranes of all living systems, not only the plasma membranes that surround cells but also those which enclose or comprise organelles such as mitochondria, vacuoles or the endoplasmic reticulum, are composed of an essentially common structure of specialised proteins and lipids. A detailed account of the structure of membranes is outside the scope of this book but the nature and properties of many of the lipid components are very important to food chemistry. Besides occurring in membranes many polar lipids occur in small amounts in crude fats and oils and are also classically associated with egg yolk where they are major constituents.

Most of the important polar lipids are *glycerophospholipids* which have two long chain fatty acid residues, conferring the lipophilic character, esterified with a glycerol molecule which also carries a phosphate group, which is of course hydrophilic in character. The phosphate group is in turn esterified with one of a number of organic bases, amino acids or alcohols which add to the hydrophilic character of the molecule. The structures of a number of glycero-phospholipids are shown in *Fig. 3.6*. Although a great many other phospholipid types have been described,

(*i*) Glycerophospholipids:

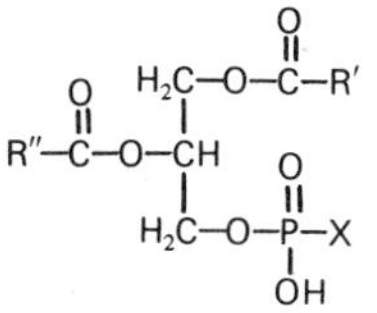

$$\begin{array}{c} R''-\overset{\overset{O}{\|}}{C}-O-CH(CH_2-O-\overset{\overset{O}{\|}}{C}-R')(CH_2-O-\overset{\overset{O}{\|}}{P}(OH)-X) \end{array}$$

R′ and R″ are the hydrocarbon chains of fatty acids typically oleic or palmitic and X is most often one of the following, to give the phospholipids shown.

$-O-CH_2CH_2\overset{+}{N}(CH_3)_3$	choline –	lecithin (phosphatidyl choline)
$-O-CH_2CH\overset{+}{N}H_3$	ethanolamine –	cephalin (phosphatidyl ethanolamine)
$-O-CH_2CH(COO^-)\overset{+}{N}H_3$	serine –	phosphatidyl serine
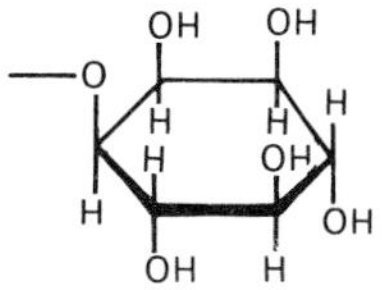 –O–(inositol ring: OH, OH, H, H, H, OH, H, OH, OH, H)	*myo*-inositol	phosphatidyl inositol

(*ii*) Cholesterol:

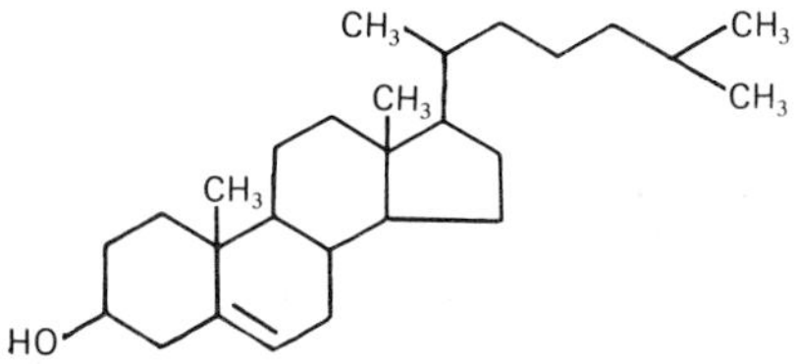

CH3 CH3 CH3 CH3 CH3 HO

FIG. 3.6. Polar lipid structures.

from the point of view of the food chemist their properties are not significantly different from those shown here. One other polar lipid that should be mentioned is cholesterol. Esters of cholesterol with fatty acids occurs as components of the lipoproteins in the blood but in membranes, and egg yolk, cholesterol is found unesterified.

Polar lipids are important in food systems because of their ability to stabilise emulsions. Emulsions are colloidal systems of two immiscible liquids, one *dispersed* in the other *continuous* phase. Foods can provide examples of both ‘oil in water’ types, *eg* milk and mayonnaise, and 'water in oil' types, *eg* butter. In many cases the structure is complicated by the presence of suspended solid particles or air forming a foam.

Emulsions are prepared by vigorous mixing of the two immiscible liquids so that small droplets of the disperse phase are formed. However, a simple emulsion usually breaks down rapidly as the dispersed droplets coalesce to form a layer which either floats to the surface or settles to the bottom of the vessel. Emulsion stability is enhanced by the presence of substances whose molecules have both polar and non-polar regions - such as phospholipids. *Figure 3.7* shows how such molecules are able to orientate themselves at the interfaces between the two phases and thus form a barrier to the coalescence of the droplets. In some food systems proteins are important emulsifying agents. Although

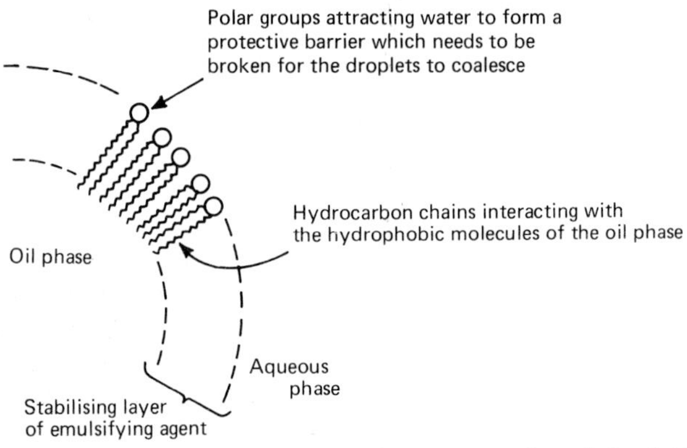

FIG. 3.7. Diagrammatic representation of the structure of an oil droplet in an oil in water emulsion.

there is little experimental evidence it seems probable that regions of the protein surface (or regions of the polypeptide chain in unfolded, denatured proteins) rich in hydrophobic amino acids (see Chapter 4) would have an affinity for the exposed surface of an oil droplet and the molecule as a whole would then take up an orientation similar to that of the polar lipid molecules in *Fig. 3.7*. The remainder of the protein molecule would then provide a hydrophilic barrier around the droplet. For fat to be digested in the small intestine it is emulsified with bile salts, derivatives of cholic acid secreted by the liver. Sodium taurocholate [3.7] is a typical example.

The traditional source of emulsifying properties in food preparation is egg yolk. Approximately 33 per cent of the yolk of a hen's egg is lipid (protein is about 16 per cent) of which about 67 per cent is triglyceride, 28 per cent is phospholipid and the remainder is mostly cholesterol. Lecithins are the predominant phospholipids with small

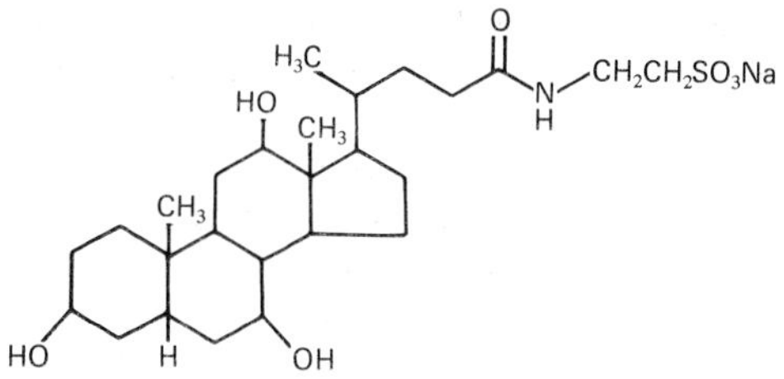

3.7

amounts of cephalins, lysophosphatidyl cholines (*ie* lecithins with only one fatty residue, on the 1 position) and sphingomyelins, [3.8].

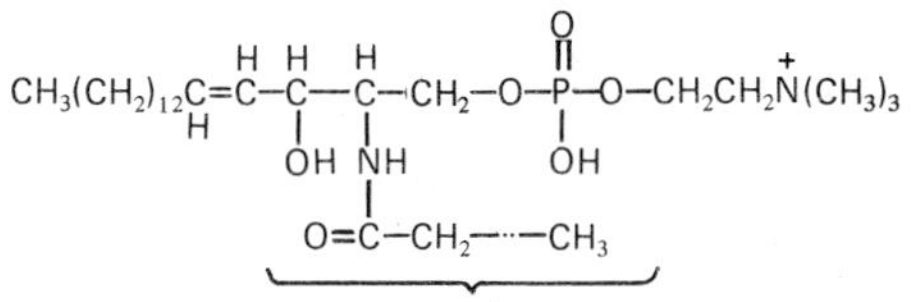

$CH_3(CH_2)_{12}CH{=}CH{-}CH(OH){-}CH(NH{-}CO{-}CH_2{\cdots}CH_3){-}CH_2{-}O{-}P(=O)(OH){-}O{-}CH_2CH_2\overset{+}{N}(CH_3)_3$

A long chain fatty acid residue

3.8

Yolk actually consists of a suspension of lipid/protein particles in a protein/water matrix. Specific associations of lipids and proteins such as these are known as lipoproteins and are employed in animal systems whenever lipid material has to be transported in an aqueous environment such as the blood. The best known applications of egg yolk as an emulsifier are in mayonnaise and sauces such as Hollandaise but its role in cake making is just the same. The lecithin recovered from soya bean oil during its preliminary purification is increasingly valued as an alternative natural emulsifying agent, particularly for use in chocolate and other confectionery, and also where a specifically vegetarian product is demanded.

Dairy products provide examples of both oil in water and water in oil emulsions. The fat in milk occurs in very stable globules mostly between 4 and 10μm in diameter. There are $1.5\text{–}3.0 \times 10^{12}$ globules in one litre of milk. Globules of such small size should, according to Stoke's Law, take some 50 hours to float to the top of a pint milk bottle and establish the familiar layer of cream but we know that in fact half an hour will suffice. This implies that much larger particles are involved with effective diameters up to 800μm. This clustering of globules is clearly no ordinary coalescence since 'evaporated milk', which is sterilised after canning by holding at temperatures above 100 °C for several minutes shows no tendency to 'cream'. The explanation for this curious behaviour lies in the nature of the 'milk fat globule membrane' which surrounds each globule. This consists largely of specific lipoproteins and polar lipids arranged around the globule in a similar arrangement to that shown in *Fig. 3.7*. The lipids of this membrane, including those associated with proteins to form lipoproteins, are typical of those found in the plasma membrane of animal cells including cholesterol and its esters with long chain fatty acids, phosphatidyl ethanolamine, choline, serine and inositol, and sphingomyelin. The membrane is most likely acquired as the milk fat globule is secreted through the plasma membrane of the mammary gland cells into the milk duct. Clustering is caused by cross-linking through one specific type of protein that occurs in the aqueous phase of the milk in very small amounts - macroglobulin. Heating above 100 °C for a few minutes (but not pasteurization) denatures this protein so that creaming is prevented.

Creaming is also prevented in homogenized milk where the fat globules are reduced to about 1 μm in diameter by passage through very small holes at high pressures and velocities around 250 m sec^{-1}. The vast increase in total surface area of the fat globules (from around 75 m^2 per litre of milk to around 400 m^2) results in extra proteins being adsorbed from the aqueous phase of the milk. These proteins prevent coalescence of the globules but do not interact strongly with the macroglobulin so that creaming no longer occurs.

Butter-making results in the formation of a water in oil emulsion. Cream with a fat content of 30–35 per cent is inoculated with a culture of bacteria and incubated for a few hours. The bacteria produce the characteristic flavour compounds such as diacetyl (butanedione). The cream is then mechanically agitated (*ie* churned) sufficiently to disrupt the fat globule membranes and cause coalescence.

During mixing a proportion of the aqueous phase, known as buttermilk, is trapped as small droplets. These are prevented from coalescence by the rigidity of the fat and by the layer of proteins and polar lipids which forms at the fat/water interfaces. Butter normally contains about 20 per cent water, to which salt is added - nowadays as a flavouring, but originally to deter the growth of unwanted micro-organisms. Margarine has buttermilk and salt added to it during the final blending stages but since it lacks its own emulsifying agents soya lecithin and other emulsifiers have to be added.

For many food products neither egg yolk nor soya lecithin are appropriate emulsifying agents and food manufacturers use synthetic substances with properties, including cost, meeting more exactly their requirements. 'Permitted emulsifiers and stabilisers'* are used in a wide range of food products including ice cream, instant desserts, cheese

---

*Stabilisers are not by themselves emulsifying agents but they enhance emulsion stability by increasing the viscosity of the aqueous phase. Most are polysaccharides such as karaya gum, locust bean gum, xanthan gum, carageenans and alginates.

(*i*) Glycerol derivatives:

$$\begin{array}{l} H_2CO\text{—}\overset{\overset{\displaystyle O}{\|}}{C}\text{—}CH_2\text{—}\cdots\text{—}CH_3 \\ \quad | \\ HCOH \\ \quad | \\ H_2COH \end{array}$$

monoglycerides

$$\begin{array}{l} H_2CO\text{—}\overset{\overset{\displaystyle O}{\|}}{C}\text{—}CH_2\text{—}\cdots\text{—}CH_3 \\ \quad | \\ HCOH \\ \quad | \\ H_2CO\text{—}\underset{\underset{\displaystyle O}{\|}}{C}\text{—}\overset{H}{\underset{OH}{C}}\text{—}CH_3 \end{array}$$

lactyl monoglycerides

Besides lactic acid derivatives monoglycerides can be linked to tartartic acid, sucrose and other similarly hydrophilic molecules.

(*ii*) Sorbitan esters:

$$\begin{array}{l} HCH \quad HC\text{—}CH_2\text{—}O\text{—}\overset{\overset{\displaystyle O}{\|}}{C}\text{—}CH_2\text{—}\cdots\text{—}CH_3 \\ HOCH \quad H\ HCOH \\ \quad\quad C \\ \quad\quad OH \end{array}$$

$$\begin{array}{l} HCH \quad HC\text{—}\overset{OH}{C}\text{—}CH_2\text{—}O\text{—}\overset{\overset{\displaystyle O}{\|}}{C}\text{—}CH_2\text{—}\cdots\text{—}CH_3 \\ HOC\text{———}COH \\ \ H \quad\quad\quad H \end{array}$$

1,5 and 1,4 sorbitan esters of fatty acids. Polyoxyethylene sorbitan fatty acid esters have one or more polyoxyethene chains:

$$\text{—}(CH_2\text{—}CH_2\text{—}O)_n\text{—}H$$

attached to the vacant hydroxyl groups of the sorbitan residue.

FIG. 3.8. Synthetic and semi-synthetic emulsifiers.

---

spread and meat products. It should not be overlooked that a major role of the protein in sausages and similar products is to bind the fat that is also included. The structures of a number of permitted emulsifying agents are shown in *Fig. 3.8*. The term 'emulsifying salts' refers to substances such as the sodium and potassium salts of phosphate, citrate and tartrate which promote emulsification by solubilising proteins which are then able to act as the emulsifying agents.

Further reading

M. I. Gurr and A. T. James, *Lipid biochemistry, an introduction*, 2nd edn. London: Chapman & Hall, 1980.

Fats and oils: chemistry and technology (R. J. Hamilton and A. Bhati, eds). London: Applied Science Publishers, 1980.

Fundamentals of dairy chemistry (B. H. Webb, A. H. Johnson and J. A. Alford, eds), 2nd edn. Westport: AVI Publishing Co, 1974.

Bailey's industrial oil and fat products (D. Swern ed), 4th edn, vol 1. New York: John Wiley and Sons, 1982.

R. J. Taylor, *The chemistry of glycerides*. Unilever Educational Booklet, Advanced Series No 4, 1968.

Egg science and technology (W. J. Stadelman and O. J. Cotterill eds), 2nd edn. Westport: AVI Publishing Co, 1977.

W. W. Christie, *Lipid analysis*, 2nd edn. London: Pergamon, 1982.

4. Proteins

The proteins are the third class of macrocomponents of living systems, and therefore foodstuffs, that we are to consider.* Proteins are polymers with molecular weights ranging from around 10 000 to several million and are usually described as having a highly complex structure. In fact there is a great deal about the structure of proteins that is quite straightforward. The monomeric units of which they are composed, the amino acids, are linked by a single type of bond, the *peptide bond* and the range of different amino acids is both strictly limited in number and essentially common to all proteins. Furthermore the 'polypeptide chain' of proteins is never branched. The special character of proteins lies in the subtlety and diversity of the variations, of both structure and function, that 'nature' works on this simple theme.

Each protein has a unique sequence of amino acids in a chain of defined length. Some proteins consist of more than one polypeptide chain held together by non-covalent forces. Except for proline and hydroxyproline all the amino acids that occur in proteins have the same general formula [4.1]:

R
|
H_2N—C—H
|
COOH

4.1

or

NH_3^+
|
H—C—COO^-
|
R

4.2

The α-carbon is asymmetrically substituted (except in the case of glycine where R = H) and as the optical configuration [4.2] shows the amino acids of proteins are members of the L-series. D-amino acids do occur in nature but not in proteins; they are found in bacteria as components of the cell wall and in certain antibiotics. Free amino acids are of little importance to food chemists although they do contri-

*In spite of their fundamental role in living systems the nucleic acids, RNA and DNA, are of almost no significance as components of our diet.

bute to the flavour of some foods, as discussed in Chapters 2 and 6. The carbonyl and amino groups in 4.2 are shown in the ionized state, known as a zwitterion, that prevails at neutral pH values. The pK_a values for the α-amino groups and carboxyl groups of amino acids are around 9.0–9.7 and 2.0–2.2 respectively.

The 20 amino acids found in most proteins are shown in *Fig. 4.1.* Two further amino acids, hydroxyproline and hydroxylysine occur in certain structural proteins including collagen (see page 93) which is of special interest to food chemists, so they are also included. There are various approaches to the classification of the amino acids but the most useful is to consider them in terms of the properties of their side chains rather than their chemical structures. The ionization of some side chains is obviously dependent on the pH of the protein's environment or the special conditions that may prevail at the active site of an enzyme.

L-Aspartic acid (asp)	$-CH_2COO^-$	Hydrophilic, acidic ($pK_a = 3.86$)
L-Glutamic acid (glu)	$-CH_2CH_2COO^-$	Hydrophilic, acidic ($pK_a = 4.25$)
L-Tyrosine (tyr)	$-CH_2-C_6H_4-OH$	Hydrophilic, neutral/acidic ($pK_a = 10.01$)
L-Cysteine (cys)	$-CH_2SH$	Hydrophilic, neutral/acidic ($pK_a = 8.33$)
L-Asparagine (asn)	$-CH_2CONH_2$	Hydrophilic, neutral
L-Glutamine (gln)	$-CH_2CH_2CONH_2$	Hydrophilic, neutral
L-Serine (ser)	$-CH_2OH$	Hydrophilic, neutral
L-Threonine (thr)	$-CH(OH)CH_3$	Hydrophilic, neutral
L-Histidine* (his)	$-CH_2$–(imidazolium ring, H–N, $\overset{+}{N}$–H)	Hydrophilic, neutral/basic ($pK_a = 6.0$)
L-Lysine* (lys)	$-CH_2CH_2CH_2CH_2NH_3^+$	Hydrophilic, basic ($pK_a = 10.53$)
L-Hydroxylysine	$-CH_2CH_2CH(OH)CH_2NH_3^+$	Hydrophilic, basic ($pK_a = 9.67$)

Amino acid	Side chain (R)	Character
L-Arginine (arg)	$-CH_2CH_2CH_2-NH-C(=\,^{+}NH_2)-NH_2$	Hydrophilic, basic ($pK_a = 12.48$)
L-Tryptophan* (trp)	$-CH_2-$ (indole ring; N, H)	Amphiphilic
Glycine (gly)	$-H$	Amphiphilic
L-Alanine (ala)	$-CH_3$	Amphiphilic
L-Phenylalanine* (phe)	$-CH_2-$ (benzene ring)	Hydrophobic
L-Valine* (val)	$-CH(CH_3)_2$	Hydrophobic
L-Leucine (leu)	$-CH_2CH(CH_3)_2$	Hydrophobic
L-Isoleucine (ile)	$-CH(CH_3)CH_2CH_3$	Hydrophobic
L-Methionine (met)	$-CH_2CH_2-S-CH_3$	Hydrophobic
L-Proline (pro)	$H_2\overset{+}{N}-\underset{}{\overset{H}{C}}-COO^-$ (ring: H_2, H_2, H_2)	Hydrophobic
L-Hydroxyproline	$H_2\overset{+}{N}-\overset{H}{C}-COO^-$ (ring: H_2, H_2, H, OH)	Hydrophilic, neutral

L-proline and L-hydroxyproline are not strictly amino acids but imino acids. The hydrocarbon side chain can be considered as being attached by both ends to give the aliphatic, cyclic structure. Essential amino acids (see page 76) are marked with an asterisk.

FIG. 4.1. The side chains (R) of the amino acids commonly found in proteins.

The formation of the peptide bond between two amino acids leads to an amide structure:

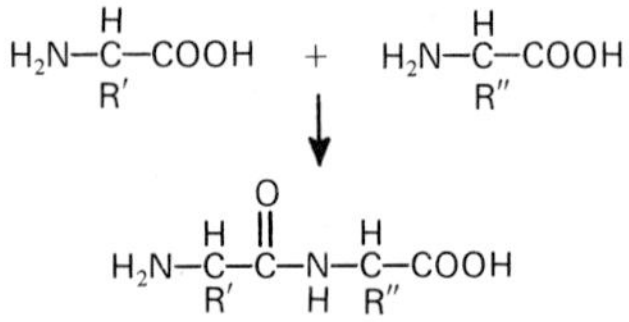

$$H_2N-\underset{R'}{\overset{H}{C}}-COOH + H_2N-\underset{R''}{\overset{H}{C}}-COOH$$

$$\downarrow$$

$$H_2N-\underset{R'}{\overset{H}{C}}-\overset{O}{\overset{\|}{C}}-\underset{H}{N}-\underset{R''}{\overset{H}{C}}-COOH$$

Of course the actual enzyme-catalysed mechanism of this reaction is considerably more involved than this simple arrow might imply. The electrons of the carbon group are delocalised to give the C-N bond considerable double bond character. As a result there is no free rotation about the C-N bond and as shown in *Fig. 4.2* all six atoms lie in a single 'amide plane'. *Figure 4.2* shows how a length of polypeptide chain can be visualised as a series of amide planes linked at the α-carbon atoms of successive amino acid residues. The polypeptide chain in *Fig. 4.2* has been drawn with scant regard for the tetrahedral distribution of the bonds about the α-carbon atoms. In reality the chain takes up more compact folding arrangements that can be defined by the angles of rotation about the C_α-N and C_α-C bonds. When a particular pair of such

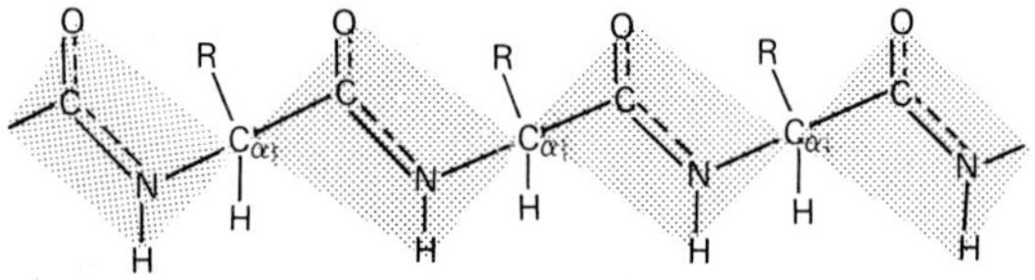

FIG. 4.2. The 'Amide Planes' of the polypeptide chain.

angles are repeated at successive α-carbon atoms then the polypeptide chain takes on a visibly regular shape which will usually be a helix but at the extreme will be a zig-zag arrangement not unlike that in *Fig. 4.2.* The folding of the polypeptide chain in a protein is determined by its amino acid sequence in ways that molecular biologists are only now beginning to understand. Within the molecule the unique patterns of folding of the peptide chain are maintained by bonds of various types. Covalent links, so-called sulphur bridges, are found linking cysteine* residues:

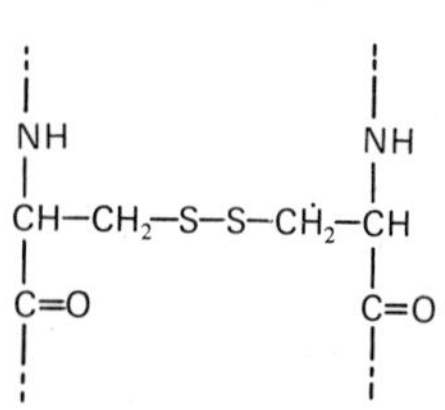

$$\begin{array}{ccc} \vdots & & \vdots \\ | & & | \\ NH & & NH \\ | & & | \\ CH & -CH_2-S-S-CH_2- & CH \\ | & & | \\ C{=}O & & C{=}O \\ | & & | \\ \vdots & & \vdots \end{array}$$

* The dimer of two cysteine molecules through their sulphydryl groups is known as cystine.

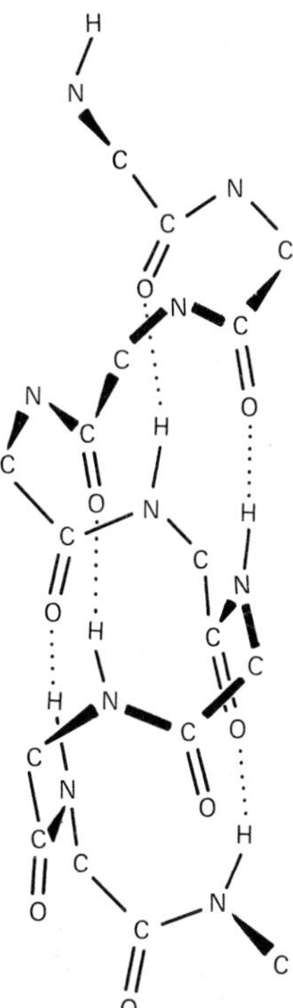

FIG. 4.3. The α-helix. Only the atoms of the backbone and those involved in hydrogen bonds (---) are shown. There are 3.6 amino acid residues per turn of the helix. Other helical structures are known with other patterns of hydrogen bonding but these do not result in such a compact structure without straining the hydrogen bonds away from their preferred linear arrangement of the four atoms: N–H---O=C.

Hydrogen bonding is specially important in maintaining ordered spatial relationships along the polypeptide chain. When these repeatedly join the carboxyl oxygen of one amino acid to the amino hydrogen of the amino acid next but three along the chain the well known α helix results as shown in *Fig. 4.3*.

The side chains of the amino acids in helical structures such as these point outwards (*ie* away from the axis of the helix) and are themselves able to form bonds that stabilise the overall folding of the polypeptide chain in globular proteins.

Cysteine–cysteine bridges have already been mentioned but inspection of *Fig. 4.1* will reveal ample scope for inter-side-chain hydrogen bonding. Hydrophobic side chains tend to be orientated so that their interaction with the aqueous environment of the protein is minimised.

Thus such residues are usually found directed towards the centre of the folded molecule where they can be amongst other non-polar residues. The tendency of these amino acids to act in this way is a major factor in the maintenance of the correct folding of the polypeptide chain and is usually referred to as 'hydrophobic bonding', a less than ideal term.

Whatever biological role a particular protein has, it is always dependent on the correct folding of the backbone to maintain the correct spatial relationships of its amino acid side chains. Not surprisingly extremes of pH and also high temperatures disrupt the forces maintaining the correct folding leading to the 'denaturation' of the protein. In a few cases highly purified proteins can be persuaded to return to their correct arrangement after denaturation but in food situations this is highly unlikely to occur. The unfolded proteins are much more likely to form new interactions with each other that lead to precipitation, solidification or gel formation. For example the white of eggs is almost entirely water (~88 per cent) and protein (~12 per cent). When egg white is heated, denaturation leads to the formation of a solid gel network in which the water is trapped. Similarly when liver is cooked the proteins contained in the liver cells are denatured and if cooking is continued for too long an unattractive hard texture results from the rigid network of polypeptide chains.

The denaturation of proteins in foodstuffs is not necessarily undesirable. Vegetables are blanched in steam or boiling water before freezing to inactivate certain enzymes, particularly lipoxygenase, which was referred to in the previous chapter.

Proteins are found in numerous roles in living systems. Enzymes, the catalysts upon which the chemical reactions that comprise all life processes depend, are proteins. The carrier molecules, such as haemoglobin, which carries oxygen in the blood, and the permeases which control the transport of substances across cell membranes, often against concentration gradients, are also proteins. Another group of proteins are the immunoglobulins which form the antibodies that provide an animal's defence against invading micro-organisms. These three classes of proteins, with enzymes by far the most numerous, are all characterised by their ability to bind specifically to other molecules as part of their physiological function. In structural terms they share a common globular arrangement of their polypeptide chains as illustrated by the diagram of myoglobin in *Fig. 4.4*. Like many enzymes and other carrier molecules myoglobin has a prosthetic group, *ie* a non-protein component, which participates in the catalytic or carrier function. In this case the iron of the porphyrin system forms a coordinate link with the oxygen molecule. It should be pointed out that the proportion of the polypeptide chain of myoglobin that is in α-helix configuration (about 75 per cent) is particularly high, most proteins have no more than 10 per cent α helix.

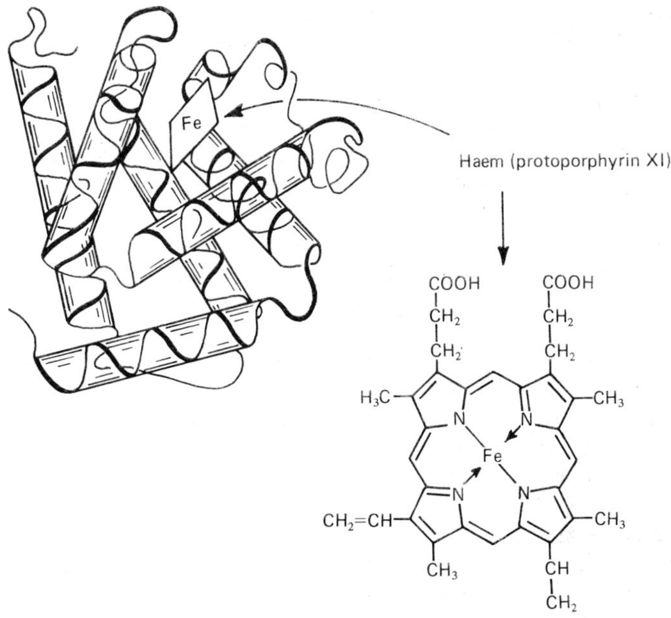

FIG. 4.4. The structure of myoglobin. Only the configuration of the polypeptide chain backbone (---C_α—N—C—C_α—N---) is shown, with regions of α helix shaded. The haem prosthetic group is contained in a 'box' lined by hydrophobic amino acid side chains.

If a protein has the same angles of rotation about the C_α-N and C_α-C bonds repeated down the length of its polypeptide chain this will obviously rule out any sort of compact, globular, folding arrangement. Instead we can expect extended, highly ordered molecules. Proteins with such a configuration have structural roles in animal tissues and one in particular, collagen, plays a crucial role in texture of meat. We will return to meat and muscle proteins later in this chapter.

The third group of proteins are those with a nutritional function, either in the transmission of nutrients from mother to offspring (the casein of milk) or the storage of nutrients to be utilized by an embryo (seed proteins of plants and egg proteins of birds). In these cases the physical characteristics of the protein will be of secondary importance to its overall chemical composition. For example gluten, the principal protein of wheat, is rich in glutamine so that compared with other proteins it has a higher nitrogen content. Amongst their other distinctive characteristics seed proteins and caseins have in common a tendency to form more or less ill-defined aggregates. Biologically such behaviour ensures that large amounts of nutrient can be concentrated without the problems of osmotic pressure usually associated with high solute concentrations. For the elements this behaviour has hindered efforts to elucidate the structure of these proteins, and as we shall see, very little is known of the structure of gluten. This is in spite of its unique ability to give us bread with an attractive crumb texture.

The protein in our diet provides the amino acids from which the body synthesises its own proteins, the major constituent of our tissues. The action of hydrolytic enzymes in the stomach as well as the small intestine breaks down food proteins to their component amino acids. On absorption into the bloodstream they then become part of the body's amino acid pool. Breakdown of the body's own tissue proteins (an essential part of the process of renewal of ageing or redundant cells) also contributes to this pool. The amino acid pool is not only drawn upon for protein synthesis but also to provide the raw materials for the synthesis of purines, pyrimidines, porphyrins and other substances.

The balancing of the pattern of amino acid supply against the needs for synthesis is a major function of the liver. Many, so-called *non-essential* amino acids, can be synthesised by mammals provided that adequate supplies of amino nitrogen and carbohydrate are available. However, there are the other, the *essential* amino acids (identified in *Fig. 4.5*) which cannot be synthesised by mammals and must be supplied in the diet.* Ideally the protein in the diet should provide the amino acids in the same relative proportions as the body's requirements but of course such an ideal is never attained. Excess supplies of particular amino acids are broken down with the carbon skeletons oxidised to provide energy or converted to fat for storage. The nitrogen is either excreted as urea or utilized in the synthesis of any non-essential amino acids that may be in short supply. Difficulties arise when it is one or more of the essential amino acids that is in short supply. Absence of even one particular amino acid will result in cessation of all protein synthesis since almost all proteins have at least one residue of all the amino acids shown in *Fig. 4.1* except hydroxyproline and hydroxylysine.

The diet of a typical European normally contains more than ample supplies of both total protein and the individual essential amino acids but for many of the people of Africa and Asia the supply of protein in the diet is inadequate in terms of both total quantity and supply of particular essential amino acids. Nutritionists have established that the proportions of the different amino acids required by the human infant correspond closely to the amino acid composition of human milk so that this is now accepted as the standard against which the nutritional values of other foodstuffs is judged. *Figure 4.5* shows the relative

---

*The rate at which humans can synthesise histidine, though adequate for adults, is insufficient to meet the demands of rapidly growing children.

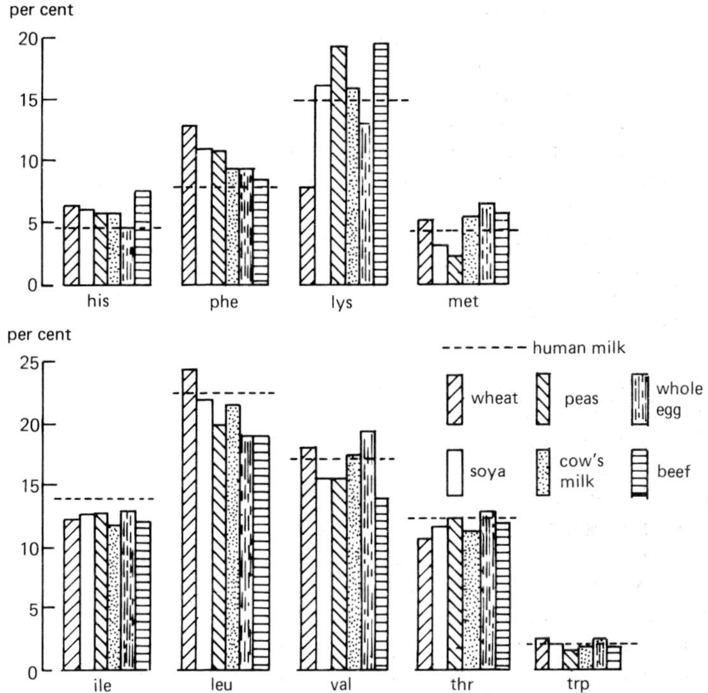

FIG. 4.5. The essential amino acids in food proteins. The relative proportions of each essential amino acid are shown expressed as the molar percentage of the total essential amino acids. The data for this figure were calculated from the data in 'First Supplement of McCance and Widdowson's *The Composition of Foods*', by A. A. Paul, D. A. T. Southgate and J. Russell, Her Majesty's Stationery Office, London.

proportions (molar rather than weight) of the essential amino acids of a number of key foodstuffs in comparison with human milk. It is immediately apparent from this data that the universality of the roles of the different amino acid side chains in the maintenance of protein structure leads to broadly similar amino acid compositions in foodstuffs of quite diverse origins. The animal protein foods, eggs, milk and meat do not differ significantly from human milk with regard to any amino acid but the plant protein foods do present some problems. The low proportion of lysine in wheat protein means that wheat is only about 50 per cent as efficient as human milk as a source of protein. An amount of wheat protein that provides just adequate levels of lysine will provide wasteful levels of the other amino acids. The legumes such as soya and peas provide ample proportions of lysine but are deficient in methionine so that they are also inefficient as sources of protein.

One should not overlook the fact that although the body utilizes these plant proteins inefficiently compared with animal proteins the greater efficiency of agricultural processes in providing plant rather than animal proteins far outweighs the utilization inefficiency. This is a consideration that should apply to the supply of food in the affluent as well as the undernourished countries of the world.

The problems posed by the low levels of lysine and methionine in what would otherwise be ideal proteins are exacerbated by the tendency of these two amino acids to undergo reactions during the storage or processing of foods which destroy their nutritional activity. The Maillard reaction between the lysine amino groups and reducing sugars is the most important and was considered in Chapter 2, but this is not the only route by which the lysine in a protein may be lost. During processing at high temperatures, especially under alkaline conditions, lysine side chains are able to form cross-links with other amino acid chains:

Serine: $HC{-}CH_2{-}OH$ or Cysteine: $HC{-}CH_2{-}SH$

$\xrightarrow{-H_2O}$ / $\xrightarrow{-H_2S}$

Dehydroalanine: $C{=}CH_2$ + Lysine: $H_2N(CH_2)_4{-}CH$

$$\downarrow$$

$$HC{-}CH_2{-}NH{-}(CH_2)_4{-}CH$$

Besides reducing the number of lysine residues available, cross-link formation of this type between neighbouring polypeptide chains will tend to prevent the assimilation of much of the remainder of the protein molecule since unfolding and access for the proteolytic enzymes of the intestine will be prevented.

The quality of a food protein in nutritional terms can only really be determined in feeding trials but sufficient is now known about protein digestion and the effects of processing techniques for fairly accurate predictions to be made, based on the total protein content and the amino acid composition. Although many reactions specific to particular amino acids have been described they have never been usefully applied to this task since they cannot provide the comprehensive, quantitative analysis that is required. Instead chemists have utilized chromatographic techniques applied to acid or enzymic hydrolysates.

Paper chromatography or thin layer chromatography with a layer of powdered cellulose will give adequate resolution of the amino acids provided that the two-dimensional technique is used. That is to say after elution by one solvent mixture (*eg* butanol : ethanoic acid : water 12 : 3 : 5 v/v/v) the paper or thin layer plate is turned through 90° for elution by a second solvent mixture (*eg* phenol : water : 0.88 ammonia 100 : 25 : 1 w/v/v). The amino acid spots are detected by means of the reaction of all amino acids (but not the imino acids) with ninhydrin, triketohydrindene hydrate, to produce a blue/violet colour:

R
H—C—NH$_2$ + 2 (ninhydrin: O, H, OH, O)
COOH

pH 5,
100°C, 15 min

(product: O, O, =N—, H, O, O) + CO_2 + RCHO

Proline and hydroxyproline produce a yellow colour.

For routine work most laboratories use ion-exchange column chromatography. After the hydrolysate has been applied to the ion-exchange resin in the column it is eluted with a series of solutions of increasing pH and ionic concentration which elute the amino acids in a highly reproducible sequence. Although a simple demonstration of the ion-exchange separation of amino acids can be performed with ordinary laboratory apparatus most laboratories use commercially manufactured 'amino acid analysers' in which all the operations of sample application, elution and the detection and estimation of the amino acids by the ninhydrin reaction as they emerge from the column, are automated. The acid hydrolysis that is required before these methods are applied inevitably converts the amide amino acids glutamine and asparagine to their corresponding acids so that usually the content of these four amino acids is reported as glutamic acid plus glutamine and aspartic acid plus asparagine. Acid hydrolysis also destroys tryptophan so that it is necessary to determine the tryptophan content of an alkaline hydrolysate by specific chemical tests.

The determination of the total protein content of a foodstuff or raw material is by no means as simple as one might imagine. Almost any chemical test, if it is to be unique to proteins, will depend on the presence of a particular amino acid side chain. Therefore analytical results will be subject to the variations in the proportions of the 'target' amino acid in the proteins of the test material and will generally need to be referred to some arbitrarily selected 'standard protein'. An example of this approach is the *Lowry method*, more popular with biochemists than food chemists as it is only applicable to proteins in solutions at concentrations up to 300 $\mu g\ cm^{-3}$. In alkaline solution, in the presence of copper ions, tyrosine residues reduce phosphomolybdotungstate to a blue compound which can be determined spectrophotometrically.

Another colorimetric/spectrophotometric procedure which is fairly uniform in its response to different proteins is the *Biuret method*. Again in alkaline conditions which ensure that the polypeptide chain is unfolded and fully accessible, Cu^{2+} ions are complexed by the nitrogen atoms of the polypeptide chain to give a characteristic violet colour:

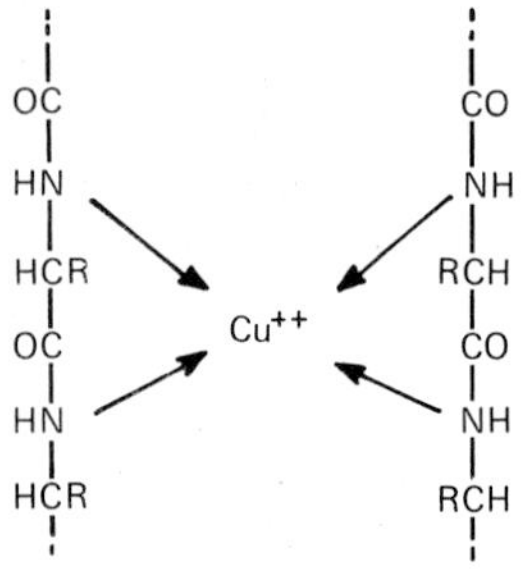

```
   OC                 CO
   |                  |
   HN                 NH
   |    ↘         ↙   |
   HCR                RCH
   |        Cu++      |
   OC                 CO
   |    ↗         ↖   |
   HN                 NH
   |                  |
   HCR                RCH
```

Potassium tartrate is included in the reaction mixture to ensure that excess Cu^{2+} ions are not precipitated as $Cu(OH)_2$.

By far the most commonly adopted method for protein determination is the *Kjeldahl procedure*. This assumes that the proportion of non-protein nitrogen in a food material is too small to be significant so that a determination of total nitrogen (excluding nitrate and nitrite which are not measured anyway) will be an accurate reflection of the total protein content. The proportion of nitrogen in most proteins is roughly 16 per cent (by weight) so that a factor of 6.25 is widely used to convert nitrogen content to protein content. Of course variations in the amino acid composition of different foods lead to the need for slightly different factors being used in accurate work with certain foodstuffs. For example cereal proteins have an unusually high proportion of glutamine resulting in a higher than usual nitrogen content and the need for a lower conversion factor, 5.70. About one third of the amino acids of gelatine are glycine which also raises the nitrogen content and a factor of 5.55 is required. Meat requires the standard 6.25 but the higher factors of 6.38 and 6.68 are required for milk and egg respectively.

In the Kjeldahl procedure the entire food sample is digested at high temperatures with boiling concentrated sulphuric acid under reflux and with a heavy metal salt (*eg* copper sulphate) present as a catalyst. A quantity of sodium sulphate is also added to raise the boiling point. Under these conditions organic material is oxidised with any organic nitrogen being retained in solution as ammonium ions. On completion of the digest an aliquot is withdrawn and transferred to an apparatus known as a Markham still. Here it is first made alkaline and then the liberated ammonia is steam distilled off into an aliquot of boric acid. The quantity of ammonia can then be obtained by titration.

Having considered some aspects of food proteins in general we can now examine the particular features of the proteins of a selection of foodstuffs.

Milk

The milk of the domestic cow, *Bos taurus*, is an important protein source for man and particularly children. Milk is an aqueous solution of proteins, lactose, minerals and certain vitamins which carries emulsified fat globules and colloidally dispersed casein micelles consisting of protein together with phosphate, citrate and calcium. If the fat is removed from milk we call the product skimmed or skim milk. If the casein is then precipitated out by reducing the pH to 4.6 (at 20 °C) the residue is known as the whey or serum.

The whey obtained in cheese making has a slightly different composition as some of the casein is solubilised and much of the lactose will have been converted to lactic acid by the action of bacteria.

The proportions of the different casein and whey proteins vary quite considerably, as does the total protein content of the milk, but the values shown in Table 4.1 are typical.

The Greek letters used in the names of the different proteins were originally assigned on the basis of the protein's mobility on electro-

Table 4.1. Typical values for the proportions of the various proteins of skimmed milk.

	Skimmed milk protein (%)		Milk ($g\,l^{-1}$)	
Casein proteins (total)	80		27	
α_s-casein		40		14
β-casein		24		8
κ-casein		12		4
γ-casein		4		1
Whey proteins (total)	20		7	
lactalbumin		12		4.2
lactoglobulin		5		1.8
Immunoglobulins		2		0.7
Others (total)		1		0.3

phoresis. The 's' of α_s-casein refers to its sensitivity to precipitation by calcium ions. Although the prefixes α_s, β *etc* define particular protein species it is now recognised that there are a number of different versions of each that differ slightly in their amino acid sequences. For example α_{s1}C-casein differs from α_{s1}B-casein (the most common variant) in having a glycine residue instead of a glutamic acid residue at position 192 in the polypeptide chain. One actually can identify the breed of cow from which a sample of milk was obtained by detailed examination of the proportions of these variants.

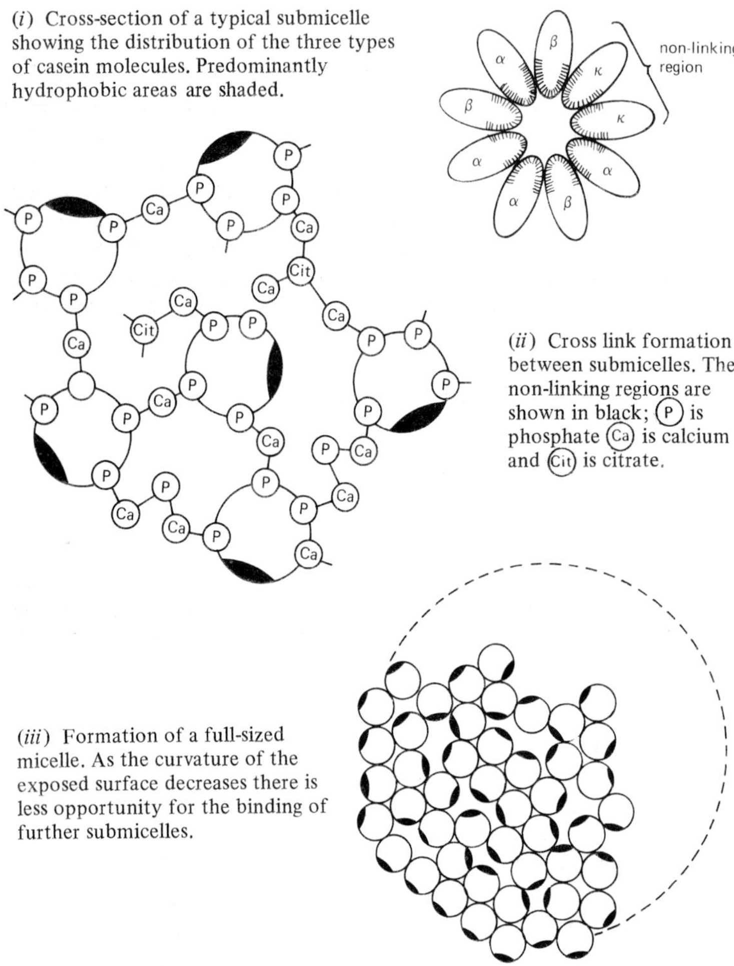

(*i*) Cross-section of a typical submicelle showing the distribution of the three types of casein molecules. Predominantly hydrophobic areas are shaded.

(*ii*) Cross link formation between submicelles. The non-linking regions are shown in black; (P) is phosphate (Ca) is calcium and (Cit) is citrate.

(*iii*) Formation of a full-sized micelle. As the curvature of the exposed surface decreases there is less opportunity for the binding of further submicelles.

FIG. 4.6. The casein micelle.

The casein micelles of milk are roughly spherical particles with diameters of 50–300 nm. Milk has about 10^{15} micelles per dm^3. A typical micelle contains some 2×10^4 casein molecules (the molecular weights of α, β and κ casein are 2.35, 2.40 and 1.90×10^4 respectively). For many years chemists have sought to describe the structural arrangement of the casein molecules within the micelle but only the very recent model described by Slattery and Evard comes near to accounting for all the observed properties of casein. It is now accepted that the micelle is an aggregate of submicelles, each consisting of 25–30 molecules of α, β and κ casein in roughly similar proportions to those in milk as a whole. γ-Casein appears to be an artefact of the procedures used to separate and purify casein molecules. It is a fragment of β-casein resulting from a limited breakdown caused by proteolytic enzymes in milk. The association of the casein molecules to form submicelles depends on the unusual characters of all three types of casein. The polypeptide chains of all three are folded into the elongated, 'rugby ball' shapes shown in *Fig. 4.6(i)*, each having, at one end, a predominance of hydrophobic amino acids. These lead to association of the molecules to form the submicelle in just the same way as the polar lipids surround an oil droplet as discussed in Chapter 3. At the outer ends there is a predominance of polar amino acids. The polar end of α_S-casein has eight serine residues to which phosphates are esterified:

The polar end of β-casein has four of these phosphoserine residues. κ-Casein differs in that it has no phosphate residues but instead one or more of the threonine residues at the polar end carry a trisaccharide unit, α-*N*-acetylneuraminyl (2→6) β-galactosyl (1→6) *N*-acetylgalactosamine [4.3].

4.3

The phosphate groups of both α_s- and β-casein react with calcium ions to link the submicelles together, either directly or in chains involving further phosphate and also citrate (see *Fig. 4.6(ii)*). The key to Slattery and Evard's proposals is the suggestion that the κ-casein molecules aggregate together in the submicelle so that their highly hydrophilic, but non-calcium binding, ends form an area, resembling a polar ice cap, where no cross linking between submicelles can occur. The increase in size as more submicelles are added (see *Fig. 4.6(iii)*) results in an inevitable tendency for the non-linking, κ-casein, areas to come to dominate the surface and ultimately prevent an indefinite increase in the size of the micelle.

The precipitation of casein to form a curd is the fundamental process involved in cheesemaking. In the case of yoghurt and some cottage cheeses the precipitation is caused entirely by low pH. The growth of *Lactobacilli* in the milk is accompanied by their fermentation of the lactose to L-lactic acid [4.4].

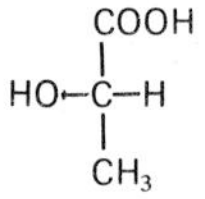

```
     COOH
      |
  HO—C—H
      |
     CH3
```

$HO\text{—}C(COOH)(CH_3)\text{—}H$

4.4

The *lactobacilli* used are sometimes those naturally present in the milk but usually a starter culture of a strain having particularly desirable characteristics is added to the milk. There is not a total precipitation of the casein at the pH of these types of fermented milk product. Associations between the casein micelles give the gel like texture that characterises yoghurt.

In the production of hard cheeses such as Cheddar, microbial action is allowed to bring the pH down to around 5.5. At this point rennet is added to bring about extensive precipitation and curd formation. Rennet is a preparation of the enzyme rennin (nowadays often called chymosin) obtained from the lining of the abomasum (the fourth stomach) of calves. In recent years the supply of calf rennin has not kept pace with demand and other, less satisfactory, enzymes have had to be used as substitutes or in mixtures with calf rennin. Proteases from fungi (mostly from *Mucor* species) and porcine pepsin are most commonly used. Rennin specifically catalyses the hydrolysis of one particular peptide bond in κ-casein as shown in *Fig. 4.7*. The κ-casein

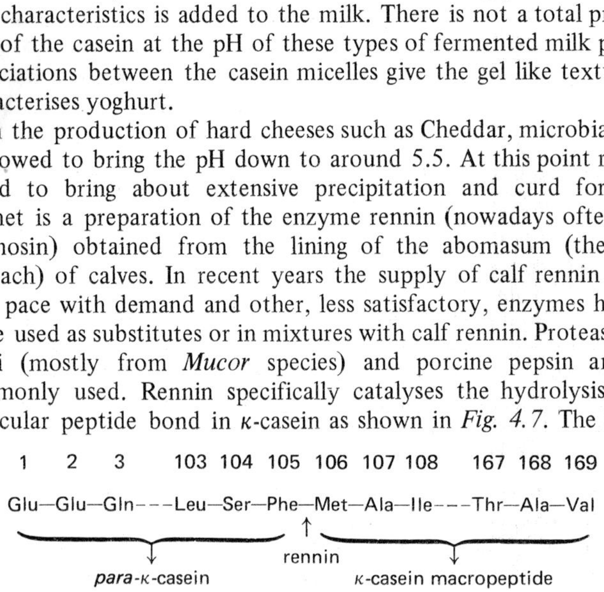

```
1   2   3     103 104 105 106 107 108     167 168 169
Glu—Glu—Gln---Leu—Ser—Phe—Met—Ala—Ile---Thr—Ala—Val
                           ↑
                         rennin
      para-κ-casein               κ-casein macropeptide
```

FIG. 4.7. The site of action of rennin on κ-casein.

is split into two fragments. One, the *para*-κ-casein remains as part of the micelle, it includes the hydrophobic section of the molecules. The other, the κ-casein macropeptide, is lost into the whey. The macropeptide fragment carries the carbohydrate units. The loss of their carbohydrate coats means that strong crosslinks between micelles can be formed and a curd rapidly develops.

The curd is held for several hours during which time the acidity increases and the curd acquires the proper degree of firmness. When this is achieved the curd is chopped into small pieces to allow the whey to run out as the curd is stirred and finally pressed. Salt is added before pressing to hasten the removal of the last of the whey and depress the growth of unwanted microorganisms. The final stage of cheese manufacture, ripening, requires low temperature storage for a considerable period. Rennet enzymes and other proteases from the microorganisms cause a limited degree of protein breakdown which is important for both the texture and flavour (many small peptides have distinctive bitter or meaty tastes) of the finished cheese. Lipases from the milk liberate short chain fatty acids from the milk fat triglycerides which also contribute to the flavour.

The special relationship between rennin action and the structure of casein obviously did not evolve to enable us to make cheese but the details of casein digestion in the calf or in man have not yet been investigated in sufficient detail to provide an explanation for the origin of the relationship.

The most abundant serum proteins α-lactalbumin and β-lactoglobulin, have attracted considerable attention from protein chemists. They have been interested in their physical properties and structure but these studies have revealed little of the reason for their presence in milk beyond their contribution to the total protein content. The immunoglobulins, in contrast, have proved to be of great interest to food scientists. The role of the macroglobulins (one class of immunoglobulins) in the creaming phenomenon was discussed in Chapter 3 but the biological role of immunoglobulins is much more important. Immunoglobulins are popularly known as *antibodies*. They are synthesised in various parts of the body (including the mammary gland) in response to the invasion of the tissues by foreign matter, particularly bacteria, viruses and toxins. Their reaction with these *antigens* is remarkably specific and facilitates the neutralisation or destruction of the invader by other parts of the body's defence mechanisms. It is now well established that exposure of a mother to many common pathogenic bacteria, especially those causing intestinal diseases such as diarrhoea which are very dangerous to the new born, leads to the appearance of appropriate antibodies in the mother's milk. These are undoubtedly effective in protecting the new born infant from many dangerous diseases. However closely the manufacturers of infant milk products based on cow's milk match the nutrient content of human milk it is becoming increasingly clear that they will not be able to match the other, unique advantages of human milk for feeding human babies.

Meat

Milk is a food that evolved purely as a food; when we consume meat or study the processes involved in the conversion of muscle tissue to the slice of roast beef on our plate we must remember that the biologist's view of its function, as well as the view of the animal itself, will be quite different from that of the food chemist or nutritionist. The structure of muscle tissue* is exceedingly complex and readers with biological or biochemical interests are strongly advised to supplement this rather rudimentary account with the details to be found in text books of cell biology, mammalian physiology or biochemistry.

A typical joint of meat from the butcher's shop is cut from a number of muscles, each having its own independent attachment to the skeleton, its own blood supply and nerves. Each muscle is surrounded by a layer of connective tissues consisting almost entirely of the protein collagen. This layer contains and supports the contractile tissues of the muscle and at its extremities provides the connections to the skeleton. The contractile units of the muscle are the muscle fibres. These are exceptionally elongated cells, 10–100 μm in diameter but often as long as 30 cm. Examined in cross section the muscle fibres can be seen to be organised into bundles separated by connective tissue which is continuous with that surrounding the muscle. Blood vessels, adipose (*ie* fatty) tissue, and nerves are also embedded in the connective tissue.

The muscle fibres have most of the features found in more typical animal cells. Surrounding the fibre is the cell membrane, known as the *sarcolemma* (the prefix 'sarco-' is from the Greek for 'flesh'). Along the length of the cell, just inside the sarcolemma, are numerous nuclei. Muscle fibres have two elaborate intracellular membrane systems, the transverse, or T, tubules, which are invaginations of the sarcolemma, and the sarcoplasmic reticulum which is the counterpart of the endoplasmic reticulum of other cells. Both participate in the transmission of the signal from the nerve endings on the surface of the fibre to the contractile elements of the fibre – the *myofibrils*. The T-tubules and sarcoplasmic reticulum wind around each myofibril so that the contraction events in each myofibril are accurately synchronised. Innumerable

---

*Although the term 'meat' usually encompasses fish and poultry as well as other mammalian tissues such as liver, kidney and intestine this account is largely restricted to the properties of the skeletal muscle tissue of mammals such as pigs, sheep and cattle.

mitochondria are also found lying between the myofibrils to ensure that the supply of adenosine triphosphate, ATP, the chemical fuel for the contraction process, is maintained.

The myofibrils are themselves composed of bundles of protein filaments arranged as shown in *Fig. 4.8(i)*. The thin filaments are composed mostly of the protein, *actin*, together with smaller amounts of *tropo-*

(*i*) *The arrangement of thick and thin filaments*

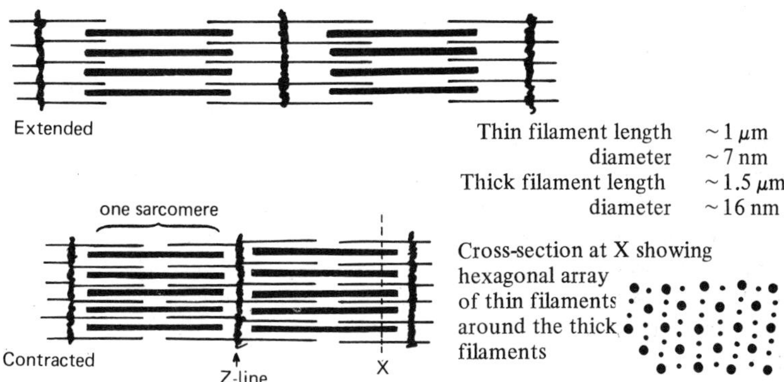

The longitudinal unit of the myofibril is known as the sarcomere. The Z-lines, which define the ends of the sarcomeres, consist of a large number of short filaments, mostly of the protein α-actinin, which link the thin filaments of neighbouring sarcomeres.

(*ii*) *The thin filament*

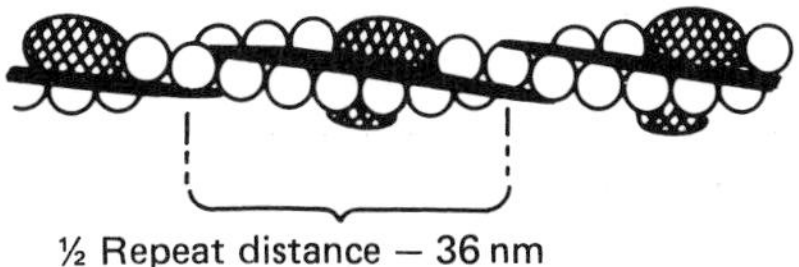

The double chain of actin molecules (closely resembling a double strand of pearls) has two strands of tropomyosin molecules lying in the grooves and troponin molecules at regular intervals.

(*iii*) *The thick filament*

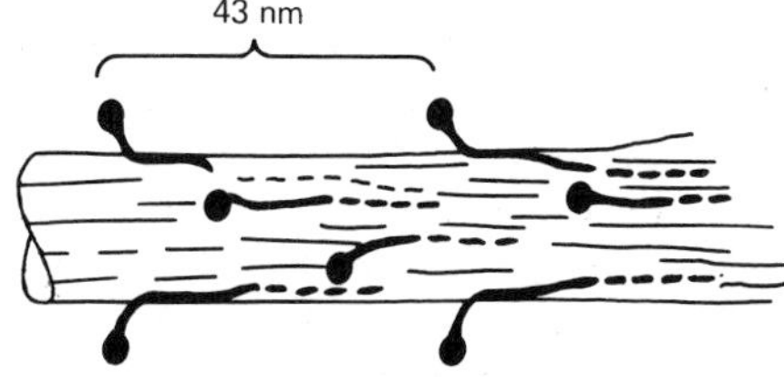

The heads of the myosin molecules protrude from the filament in six lines which are orientated to interact with the actin molecules of the surrounding thin filaments. At the centre of the filament the myosin molecules change direction resulting in a region lacking heads. The myosin molecule is about 160 nm long.

FIG. 4.8. The myofilaments of skeletal muscle.

myosin and *troponin* as shown in *Fig. 4.8(ii)*. The thick filaments are aggregates of the very large (molecular weight 5×10^5) elongated protein, *myosin*, arranged as shown in *Fig. 4.8(iii)*. It is now well established that the sequence of events in muscle contraction is as follows:

(*i*) The nerve impulse is transmitted throughout the cell by the T-tubules and causes the release of Ca^{2+} ions from the vesicles which the membranes of the sarcoplasmic reticulum form.
(*ii*) The Ca^{2+} ions bind to the troponin of the thin filaments. This causes a change in the shape of the troponin molecules which moves the adjoining tropomyosin. The change in the position of the tropomyosin exposes the 'active site' of the actin molecules.
(*iii*) The 'activated' actin molecules are now able to react repeatedly with the myosin of the thick filament and ATP, the energy source for contraction:

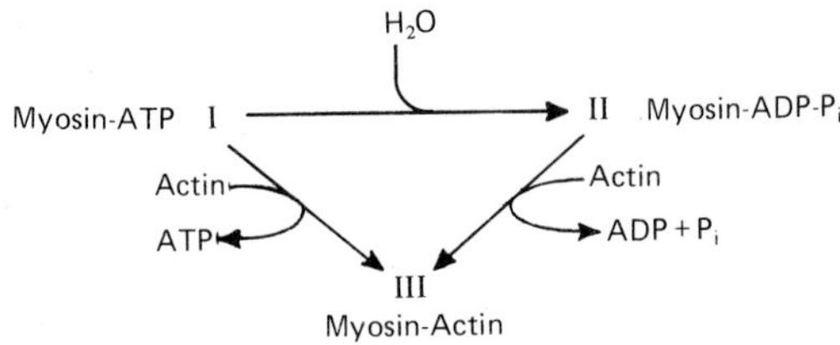

(*iv*) The heads of the myosin molecules are at different angles to the thick filaments at different stages in this cycle so that with each turn of the cycle, and with the hydrolysis of one ATP molecule, the myosin head engages with a different actin molecule and the two filaments move with respect to each other and contraction occurs.
(*v*) When the nerve impulse ceases, calcium is pumped back into the sarcoplasmic reticulum and the cycle comes to rest at stage II as the tropomyosin returns to its original position and the actin can no longer interact with a myosin head. The muscle is now relaxed and will be returned to its original extended state by the action of other muscles.

The energy, as ATP, for contraction and the active uptake of calcium by the sarcoplasmic reticulum, is derived from two sources. When moderate levels of muscle activity are demanded pyruvic acid (2-oxopropanoic acid) [4.5] derived from carbohydrate breakdown or acetyl units derived from fatty acid breakdown, is oxidised in the mito-

$$\begin{array}{c} CH_3 \\ | \\ C=O \\ | \\ COOH \end{array}$$

4.5

chondria of the muscle fibres. The energy made available by this oxidation is utilized in the mitochondria for the phosphorylation of adenosine diphosphate to the triphosphate. Three molecules of ATP are formed for every $\frac{1}{2}O_2$ reduced to H_2O. The oxygen is carried from the blood to the mitochondria as the oxygen adduct of myoglobin, the source of the red colour of fresh meat. When short bursts of extreme muscle activity are demanded, for example when one runs for a bus or when a bird like a pheasant briefly takes to the air the limited rate of oxygen transport would not give an adequate supply of ATP. At these times the muscle obtains its energy by the less efficient, but anaerobic and more rapid conversion of glycogen or glucose to lactic acid by the metabolic process of *glycolysis*. Only two molecules of ATP per molecule of glucose are obtained in glycolysis compared with 36 when the glucose is oxidised completely to CO_2. In the muscles of the living animal anaerobic glycolysis cannot be maintained for many seconds before there is an unacceptable accumulation of lactic acid but as soon as demands on the muscles ease the lactic acid is readily oxidised back to pyruvate which can then be either oxidised to CO_2 or some can be reconverted back to glucose in the liver.

On the death of the animal there is still a demand for ATP in the muscle from the pumping systems of the sarcoplasmic reticulum and other reactions even though the muscles are no longer being called upon to contract by signals from the nervous system. As there is obviously no further supply of oxygen from the blood stream the supply of ATP is maintained by glycolysis for some time until one of two possible circumstances bring it to a halt. If the animal was starved and otherwise badly treated before slaughter the muscle's reserves of glycogen will be rapidly exhausted and glycolysis will cease. However, if the animal's muscles did have adequate glycogen reserves glycolysis will still cease after a few hours when the accumulation of lactic acid lowers the pH to around 5.0–5.5, sufficient to inhibit the activity of the glycolytic enzymes. When the ATP level in the muscle has consequently fallen calcium will no longer be pumped out of the sarcoplasm into the vesicles of the sarcoplasmic reticulum. This calcium allows the myosin to interact with the actin but the lack of ATP will prevent the operation of the full cycle which would normally lead to contraction. The thick and thin filaments will therefore form a permanent link at stage III and the muscle will go rigid - the state of *rigor mortis*. It is clearly impossible to get meat cooked before it goes into *rigor* and meat cooked during *rigor* is said to be exceedingly tough.

The answer to the problem of *rigor mortis*, and, as we shall see other aspects of meat quality, lies in butchers ensuring that the critical pH range of 5.0–5.5 is achieved. (Suffice to say it is the application of sound pre- and post-slaughter practices rather than the possession of a pH meter that makes a good butcher.) The most obvious effect of the low pH is to deter the growth of the putrefactive and pathogenic micro-organisms that spread from the hide, intestines *etc* during the preparation of the carcass. During the conditioning (*ie* hanging) of the carcass the toughness due to *rigor mortis* disappears. A highly specific protease, active only around pH 5 in the presence of Ca^{2+} ions, has been detected in muscle tissue. What use it is to the live animal is total obscure but in meat it catalyses the breakdown of the thin filaments at the point where they join the Z-line filaments. This creates planes of weakness across the muscle fibres which ensure that properly conditioned meat is suitably tender.

Muscle tissue contains a great deal of water, 55–80 per cent. Some of this water is directly bound to the proteins of the muscle, both the myofilaments and the enzyme proteins of the sarcoplasm. The bulk of the water, however, occupies the spaces between the filaments which are too narrow to contain dissolved proteins. Protein denaturation on cooking releases some of this water to give meat its desirable moistness. The housewife likes the cut surface of a joint of meat to look moist and shiny in the raw state. Although the sarcoplasmic proteins will bind less water at pH values near to their isoelectric points, which do happen to be mostly around pH 5, the principal cause of moisture release, drip, from well-conditioned meat, is that as the pH falls the gap between neighbouring filaments is reduced and some water is squeezed out. Although the butcher would obviously not wish too much moisture to be lost from a joint before it is weighed and sold, the increased mobility of the muscle water has another beneficial effect besides giving the meat an attractive moist appearance.

The further effect is, as we shall see, on the colour of the meat. It has already been mentioned that the red colour is due to the presence of myoglobin (see *Fig. 4.4*). The amount of myoglobin varies from muscle to muscle, from species to species and with the age of the animal. Some typical data for cattle and pigs are given in Table 4.2 (based on results quoted by R. A. Lawrie, 1979 - see 'Further

Table 4.2. The myoglobin content of different muscles.

Animal	Age	Muscle	Myoglobin content/ percentage of fresh weight
Calf[1]	12 days	*Longissimus dorsi*[6]	0.07
Steer[2]	3 years	*Longissimus dorsi*[6]	0.46
Pig[3]	5 months[3]	*Longissimus dorsi*[6]	0.030
Pig[4]	7 months[4]	*Longissimus dorsi*[6]	0.044
Pig[5]	7 months[5]	*Rectus femoris*[7]	0.086

1, Veal; 2, beef; 3, pork; 4, bacon; 5, pork sausage; 6, the principal muscle of 'rib of beef' or 'short back bacon'; 7, the principal muscle of 'leg of pork'.

reading', p. 100). As a general rule muscle used for intermittent bursts of activity, which will be fuelled predominantly by anaerobic glycolysis, has only low levels of myoglobin, few mitochondria and will thus be pale coloured. The breast, *ie* the flight muscle, of poultry is an extreme example. In contrast muscles used more or less continuously, such as those in the legs of poultry, will be fuelled by oxidative reactions and will consequently require high levels of myoglobins and be much darker.

The state of the iron in myoglobin also affects its colour. As shown in *Fig. 4.4* four of the six coordination positions of the iron are occupied by nitrogen atoms of the porphyrin ring. The fifth, perpendicular to the plane of the ring, is occupied by the nitrogen atom of a histidine in the polypeptide chain. The sixth position, opposite to the fifth, is where oxygen is bound, as shown in *Fig. 4.9*. In the live animal

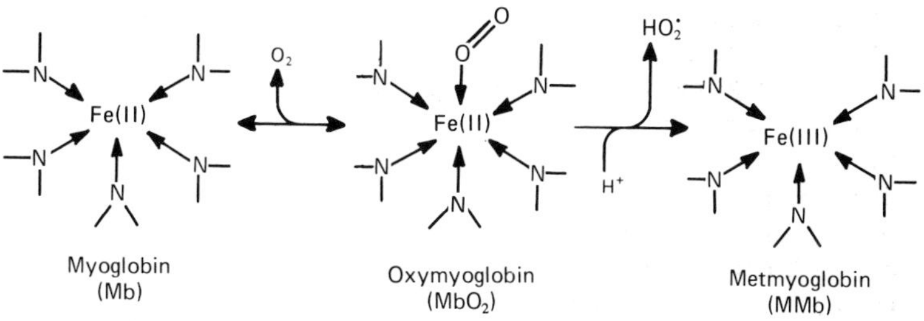

Myoglobin (Mb) — Oxymyoglobin (MbO_2) — Metmyoglobin (MMb)

FIG. 4.9. Oxygen binding by myoglobin.

the iron remains in the reduced, Fe(II) state. Oxymyoglobin is bright red in colour (its spectrum has a double peak, λ_{max} values 544 and 581 nm) whereas myoglobin is a dark purplish red (λ_{max} 558).

The interior of the muscle tissue of a carcass will obviously be anaerobic and the freshly cut surface of a piece of raw meat is, as one would expect, the corresponding shade of dark red/purple. Soon, diffusion of oxygen into the surface begins to convert myoglobin (Mb) to oxymyoglobin (MbO_2) and the meat acquires an attractive bright red appearance as a layer of MbO_2 develops. The importance of an adequate *post-mortem* pH fall is that it is the free moisture of the meat through which the oxygen diffuses. The respiratory systems of the muscle continue to consume any available oxygen until the meat is actually cooked so that the MbO_2 layer is never more than a few millimetres deep. At low oxygen concentrations, as prevail at the interface between the MbO_2 and Mb layers the iron of the myoglobin is oxidised by the oxygen it binds to convert it to the Fe(III) state as shown in *Fig. 4.9*. This has two consequences. The most obvious is that the metmyoglobin (MMb) which is produced is a most unattractive shade of brown (λ_{max} 507 nm). Meat kept at room temperature therefore develops a brown layer just below the surface and as time goes on this widens until the surface of the meat looks brown rather than red - a sure sign to the housewife that the meat is no longer fresh. The other consequence is that the oxygen departs as a free radical, the protonated form of the superoxide anion. With the exception of the bacteria that are obligate anaerobes all living organisms have a pair of enzymes, superoxide dismutase (SODM) and catalase, whose role is to destroy this dangerous by-product of oxygen metabolism:

$$2HO_2^{\cdot} \xrightarrow{2H^+} 2O_2^{\cdot-} \xrightarrow[\text{SODM}]{O_2} \underset{\text{peroxide anion}}{O_2^{2-}} \xrightarrow[\text{CATALASE}]{H^+} OH^- + \tfrac{1}{2}[O_2]$$

In spite of the activity of these two enzymes the free radical may well have time to initiate the chain reactions of fatty acid autoxidation that lead to the rancidity of the fatty parts of the meat. Thus there is a clear association between the appearance of the raw meat and its flavour when cooked.

When meat is cooked the myoglobin is denatured along with most other proteins. The unfolding of the polypeptide chain displaces the histidine whose nitrogen atom was linked to the iron of the porphyrin ring. This changes the properties of the iron so that it is now readily oxidised to the Fe(III) state by any oxygen present. The result is that except for the anaerobic centre of a large joint the meat turns brown. The pink colour at the centre of a roast joint is due to a haem derivative in which histidine residues (perhaps, but not necessarily, from the original myoglobin molecule) occupy both of the vacant coordination positions on opposite faces of the porphyrin ring.

Salting has always been an important method of meat preservation. Rubbing the carcass with dry salt, steeping in brine or modern injection methods all aim to raise the salt concentration in the tissue and thereby inhibit microbial growth. In the case of bacon, low levels of nitrites are traditionally included with the salt. Under the influence of enzyme action in the meat the nitrite gives rise to nitrogen oxide (NO) which combines with the myoglobin to form a red pigment, nitrosyl myoglobin (MbNO). This gives uncooked bacon and ham its characteristic red colour. When bacon or ham is cooked denaturation of the MbNO leads to the formation of the bright pink pigment usually known as nitrosyl haemochromogen. This is a porphyrin derivative of uncertain structure but probably having two nitroso- groups bound to the iron atom.

Although the *post-mortem* behaviour of the myofibrillar structures is an important factor in the texture of meat the consumer is much more aware of differences in texture that stem from the maturity of the animal and of course the part of the carcass the meat came from. These variations arise in the connective tissue of the muscle rather than the contractile tissues. The connective tissue of the muscle is composed almost entirely of fibres of the protein *collagen*. Collagen fibres consist of cross linked, longitudinally arranged *tropocollagen* molecules arranged as shown in *Fig. 4.10(i)*. Each tropocollagen molecule is 280-300 nm in length. The orderly array of these molecules generates a pattern of transverse bands across the fibre with a repeat distance of 64 nm. The tropocollagen molecule itself consists of three very similar polypeptide chains. The conformation of each polypeptide chain is an extended left-handed helix (remember that the classic α-helix shown in *Fig. 4.3* is a compact right-handed helix) and the three chains are then wound together in a right-handed helix, *Fig. 4.10(ii)*. The amino acid composition and sequence of tropocollagen is striking. Roughly one third of all the amino acid residues are glycine, and proline plus hydroxyproline account for between 20 and 25 per cent. The sequence given in *Fig. 4.10(iii)* shows that in fact glycine occurs at every third residue along the chain. Structural studies show that the glycine

(*i*) *The arrangement of tropocollagen molecules in the fibre*

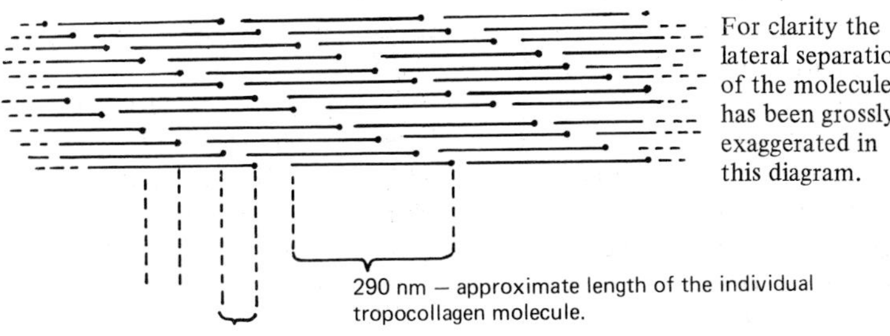

(*ii*) *The three polypeptide chains of tropocollagen*

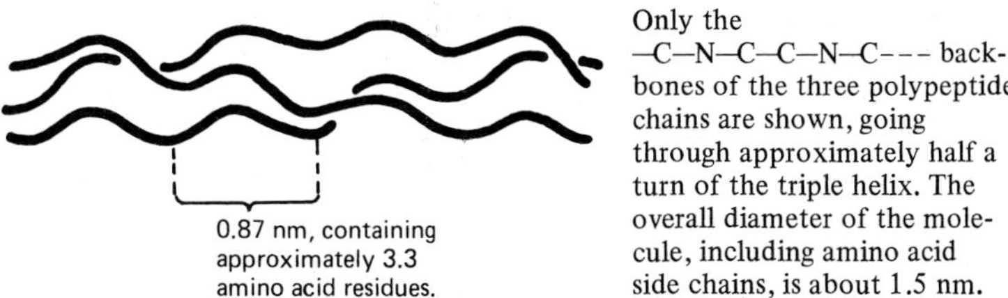

(*iii*) *The amino acid sequences of a section of a tropocollagen polypeptide chain:*

—Ser—Gly—Pro—Arg—Gly—Leu—Hyp—Gly—Pro—Hyp—Gly—Ala—Hyp—Gly—

FIG. 4.10. The molecule structure of collagen.

residues always occur at the points in the helical structure where the three chains approach closely. Only the minimal side chain of glycine - a solitary hydrogen atom - can be accommodated in the small space available. The two imino acids are important since their particular form of the peptide bond gives the ideal bond angles for the tropocollagen helix. The structure is maintained by hydrogen bonding between the three chains and also by hydrogen bonds involving the hydroxyproline hydroxyl group. The hydroxylation of proline to hydroxyproline occurs after the protein has been synthesised. The observation that ascorbic acid, vitamin C, is the reducing agent in the reaction can be readily correlated with the clinical signs of the deficiency of the vitamin discussed in Chapter 7. However strong an individual tropocollagen molecule is, the strength of the fibre as a whole will depend on lateral cross links to ensure that neighbouring molecules do not slide over each other under tension. A number of types of cross links have been identified; one of the simplest and most abundant types is shown in *Fig. 4.11*. Cross links are most abundant in connective tissues where greatest strength is required, such as the Achilles tendon.

It is a common observation that meat from young animals is more tender than that from older ones. It has now been realised that this is due to an increase, as the animal ages, in the number of cross links rather than an increase in the proportion of total connective tissue. There are, of course, variations in the amount of collagen between different muscles of the same animal which are very important to meat texture. The most tender cuts, and the most expensive, are those with the lowest proportion of connective tissue. For example in beef the connective tissue content of the *Psoas major* muscle (fillet steak) is one third of that in the *Triceps brachii* muscle (stewing beef).

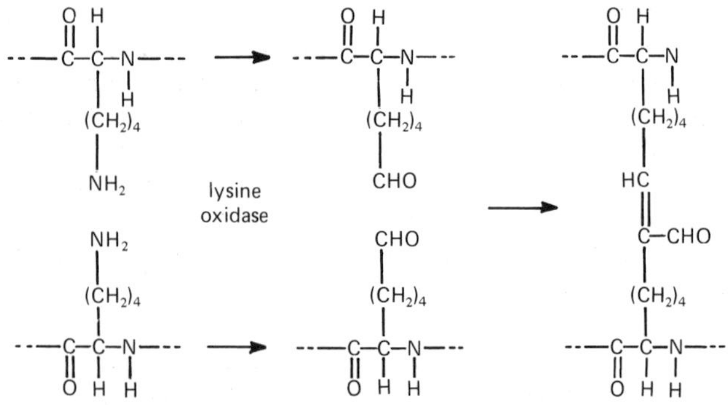

FIG. 4.11. Cross link formation in collagen between two lysine residues.

When meat is heated the hydrogen bonds that maintain collagen's structure are weakened. Very often the fibres shorten as the polypeptide chains adopt a more compact helical structure. If the heating is prolonged, as in a casserole, not only hydrogen bonds but also some of the more labile cross links will be broken. The result will be solubilisation of the collagen, some of which will be leached out and cause gelatinisation of the gravy as it cools and the hydrogen bonds are re-established.

Bread

Bread has been eaten in the temperate zones of the world for thousands of years. It is only in the last few that chemists and biochemists have been able to address themselves seriously to the fundamental problem of bread. That is: why, of all the cereals, wheat, rye, barley, oats, sorghum, maize and rice, it is only the first two that will give us bread with a leavened, open crumb structure.

The grains (*ie* the seeds) of cereals have a great deal in common. They consist of three major structures, (*i*) the embryo or germ of the new plant, (*ii*) the endosperm which is the store of nutrients for the germinating plant, and (*iii*) the protective layers of the seed coat which are regarded as bran by the miller. The endosperm is about 80 per cent of the bulk of grain. White flour is almost pure endosperm and since we are primarily interested in bread what follows refers to the endosperm and endosperm constituents of wheat.* The endosperm consists of tightly packed thin walled cells of variable size and shape. The cell walls are the origin of the small proportions of cellulose and hemicelluloses in white flour. The cells are packed with starch granules lying in a matrix of protein. The protein, some 7–15 per cent of the flour, is of two types. One type (about 15 per cent of the total) consists of the residues of the typical cytoplasmic proteins, mostly enzymes, which are soluble in water or dilute salt solutions. The remaining 85 per cent are the storage proteins of the seed, insoluble in ordinary aqueous media and responsible for dough formation. These dough forming proteins are collectively referred to as *gluten*. The gluten can be readily extracted from a flour by adding enough water to form a dough, leaving the dough to stand for half an hour or so, and then finally kneading the dough under a stream of cold water which washes out all the soluble material and the starch granules. The resulting tough, visco-elastic and sticky material contains about one third protein and two thirds water.

---

*All the common bread wheats in cultivation are treated by plant taxonomists as subspecies of the single species *Triticum aestivum*.

Flours from different wheat varieties vary in protein content. In general, flours that are good for breadmaking (*ie* give a good loaf volume) are given by the spring sown wheat varieties grown in North America. These varieties tend to have higher protein contents (12–14 per cent). Good breadmaking wheats are most often the type described as 'hard' by the miller, *ie* the endosperm is brittle and disintegrates readily on milling. The baker's description of a good breadmaking flour as 'strong' refers to the characteristics of the dough – it is more elastic and more resistant to stretching than the dough of a 'weak' flour. Weak flours are essential for biscuits and short-crust pastry. These flours are usually obtained from the 'soft' winter wheats (actually sown in the autumn!) that are grown in Europe. Their protein content is usually less than 10 per cent. In spite of the general correlations between protein content and milling/baking properties there are enough exceptions to demonstrate that it is the properties of different wheat proteins rather than their abundance that is the major factor.

The gluten proteins can be fractionated on the basis of their solubilities. The most soluble, the *gliadins*, can be extracted into 70 per cent ethanol. The gliadins constitute about one third of the gluten. The remaining two thirds are the *glutenins* which are extremely difficult to dissolve fully. A solution consisting of 0.1 M ethanoic acid, 3 M urea and 0.01 M cetyltrimethylammonium bromide (cetrimide) will dissolve all but 5 per cent of the glutenins. What is known of the molecular structure of the gluten proteins does go some way towards explaining their remarkable dough forming and solubility characteristics. Using elaborate electrophoretic techniques it has been shown that a single variety of wheat may have over 40 different gliadin proteins. Most have molecular weights of 30–40 000 and though they all differ slightly in amino acid composition they have much in common. Their contents of glutamine (36–45 mole per cent) and proline (15–30 mole per cent) are exceptionally high compared with other proteins. These two amino acids contain an above average proportion of nitrogen and are also particularly easily utilised by the germinating seed as sources of both nitrogen and carbon. The less easily utilised nitrogen-rich amino acids (arginine, lysine and histidine) occur in unusually small proportions. The gliadins also contain rather low levels of aspartate and glutamate so that there is an overall scarcity of charged amino acids. As a result the gliadins have very little ionic character compared with other proteins – a clear explanation of their lack of solubility. Complete amino acid sequences for purified gliadins have not yet been obtained but sequences of *N*-terminal fragments show many similarities. For example seven different gliadins from the wheat variety Ponca have an essentially common sequence of 25 amino acids at their *N*-terminal ends.

The glutenins have proved much more difficult to characterise. Their molecular weights range from 40 000 to 20 million with most around 2 million. When glutenins are treated with reagents such as mercaptoethanol which break disulphide bridges (*ie* those between cysteine sulphydryl groups) a range of about 15 different protein subunits is obtained, with molecular weight from 11 000 to 133 000 but mostly around 45 000. These subunits are not apparently identical with any known gliadins although they do have broadly similar amino acid compositions to the gliadins, *ie* about one third glutamine and about one tenth each of proline and glycine. One particular glutenin subunit, admittedly an extreme example, has these three amino acids together constituting almost three quarters of the total. It is now believed that the glutenins consist of linear chains of these subunits. It is unlikely that more than one disulphide bridge is involved in each intersubunit link. The other few cysteine residues in each subunit are available for intrasubunit links which help to maintain the tightly coiled arrangement of the polypeptide chain in the subunit.

When water is added to flour a dough is formed as the gluten proteins hydrate. Some water is also, of course, taken up by the damaged starch granules. The viscoelastic properties of the dough depend upon the glutenin fraction which is able to form an extended three-dimensional network. Links between the glutenin chains depend on different types of bonding. Hydrogen bonding between the abundant glutamine amide groups is probably most important but ionic bonding and hydrophobic interactions also have a role. The gliadin molecules are assumed to have a modifying influence on the viscoelastic properties of the dough. It is now well established that the relative proportions of the high and low molecular weight proteins, *ie* the glutenins and gliadins, is a major factor in the breadmaking character of a flour. In general a higher proportion of glutenins results in doughs that are stronger, require more mixing, and gives loaves of greater volume. It is not always fully appreciated that bread is *sold* by weight, loaves having to conform to certain standard weights, but in the shop a loaf will be *bought* on the basis of its volume. For biscuits (except crackers) and many types of cakes and pastry a good breakmaking flour is most unsuitable. Made with a strong flour, biscuits would be hard rather than crisp and would tend to shrink erratically after moulding. These, and similar problems with pastry making, do not cause difficulty in the home but obviously a mechanised commercial bakery will need to control very closely the properties of the flour it uses.

The breadmaking properties of a flour are much improved by prolonged storage. Autoxidation of the polyunsaturated fatty acids of flour lipids results in the formation of hydroperoxides which are powerful oxidising agents. One consequence is a bleaching of the carotenoids in the flour giving the bread a more attractive, whiter crumb. However, the most important of the beneficial effects of ageing are on the loaf volume and crumb texture. Over the first twelve months of storage of flour there is a steady increase in the loaf volume and the crumb becomes finer and softer. Longer storage results in worsening baking properties. For over 50 years it has been common practice to simulate the ageing process by the use of oxidising agents as flour treatments, at the mill or as additives at the bakery. The agents used have included chlorine dioxide (applied as a gaseous treatment by the miller), benzoyl peroxide (di(benzenecarbonyl) peroxide), ammonium and potassium persulphates, potassium bromate and iodate and azodicarbonamide [4.6]. Chlorine dioxide and benzoyl peroxide are usually

$$H_2N-\overset{O}{\overset{\|}{C}}-N=N-\overset{O}{\overset{\|}{C}}-NH_2$$

4.6

described as bleaching agents and the others as flour improvers but there is no clear distinction between the two types of activity. Most are used at levels between 10 and 100 mg kg^{-1} of flour. Another improving agent of increasing importance in modern bakery practice, ascorbic acid, which will be discussed below, is actually a reducing agent rather than an oxidising agent.

A satisfactory dough is one which will accommodate a large quantity of gas and retain it as the protein 'sets' during baking. The achievement of such a dough requires more than the simple mixing of the ingredients – mechanical work has also to be applied. In traditional breadmaking the kneading of the dough provides some of this work but the remainder is performed by the expanding bubbles of the carbon dioxide that is evolved by the yeast fermentation. At the molecular level these processes, collectively referred to as dough development, are not well understood and the account of them that follows is unlikely to represent the final word on the subject.

It is generally believed that during development the giant glutenin molecules are stretched out into linear chains which interact to form elastic sheets around the gas bubbles. A number of chemical reactions are involved in this. The mechanical stresses are sufficient to break, temporarily, the hydrogen bonds which are important in binding together the different gluten proteins. Other reactions involve the sulphydryl groups of the proteins. Under mechanical stresses exchange reactions between neighbouring sulphydryl groups will allow the glutenin subunits to take up more extended arrangements. These exchange reactions (*Fig. 4.12*) require the presence of low molecular weight sulphydryl compounds such as glutathione (GSH) [4.7] which has been found in flour in sufficient quantities (10–50 mg kg^{-1} of flour).

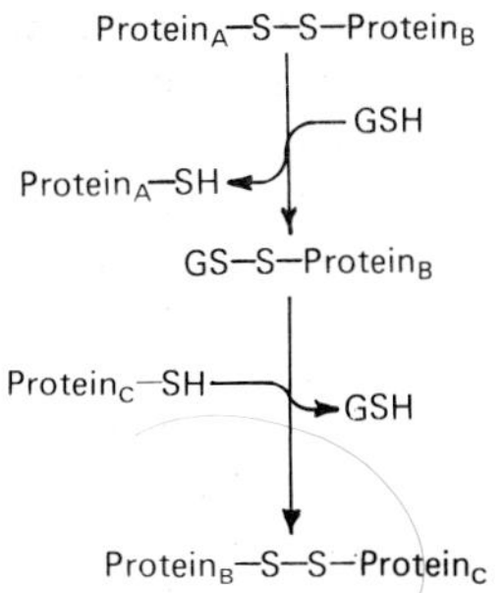

$$\text{Protein}_A\text{–S–S–Protein}_B \xrightarrow[\text{Protein}_A\text{–SH}]{\text{GSH}} \text{GS–S–Protein}_B \xrightarrow[\text{GSH}]{\text{Protein}_C\text{–SH}} \text{Protein}_B\text{–S–S–Protein}_C$$

FIG. 4.12. Disulphide exchange reactions in dough. (Based on N. Chamberlain (p. 87) in *Vitamin C: ascorbic acid* (J. N. Counsell and D. H. Hornig eds), London: Applied Science, 1981.)

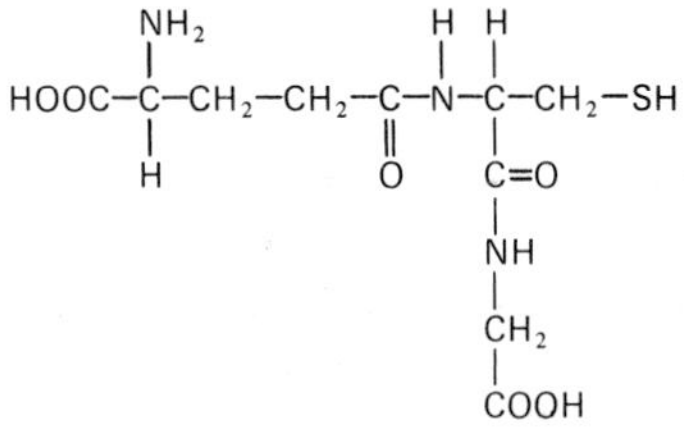

$$HOOC{-}CH(NH_2){-}CH_2{-}CH_2{-}C({=}O){-}NH{-}CH(CH_2{-}SH){-}C({=}O){-}NH{-}CH_2{-}COOH$$

4.7

The effect of natural ageing, and agents such as bromate (the most studied oxidising improver) on dough rheology and loaf volume is through the oxidation of free sulphydryl groups in the flour. With the correct degree of oxidation the extent of disulphide exchange will be just sufficient to permit optimum expansion without the dough becoming excessively weak. It seems likely that bromate acts directly on the free sulphydryl groups of the glutenin subunits - oxidising either pairs of neighbouring sulphydryls to give disulphides:

$$-SH + HS- \longrightarrow -S-S-$$

or single sulphydryls to sulphonates:

$$-SH \longrightarrow -SO_3^-$$

Traditional baking processes require prolonged periods (up to 3 hours) of fermentation before the dough is ready for the oven and a great deal of research has taken place to find ways of accelerating the dough development process. After many attempts in both Britain and the US it was at the laboratories of the Flour Milling and Baking Research Association (at Chorleywood, Hertfordshire) that the first really successful process was devised. Known internationally as the Chorleywood Bread Process (CBP) it depends upon the use of ascorbic acid (AA) as an improver coupled with high speed mixing of the dough. When the flour, water and other ingredients are mixed the ascorbic acid is oxidised to dehydroascorbic acid (DHAA) (see Chapter 7) by ascorbic acid oxidase, an enzyme naturally present in flour:

$$AA + \tfrac{1}{2}O_2 \longrightarrow DHAA + H_2O$$

The dehydroascorbic acid, an oxidising agent, effects its improving action by means of a second enzyme, also naturally present in flour, which catalyses the conversion of glutathione to its oxidised dimeric form:

$$DHAA + 2GSH \longrightarrow AA + GSSG$$

which is of course inactive in the disulphide exchange reactions of dough development. The elimination of long fermentation periods has made the CBP very popular with bakers, even though it requires special mixers capable of 400 rpm and monitoring of the energy imparted to the dough (5 watt-hours per lb of flour (4×10^4 J kg^{-1}) is required). The resulting bread need not be easily distinguishable from that produced by traditional, bulk fermentation methods. However, it is often observed that the flavour of CBP bread is inferior - this is apparently because the yeast has had less time in which to contribute ethanol and other metabolic by-products to the dough.

Further reading

R. E. Dickerson and I. Geis, *The structure and action of proteins*. California: W. A. Benjamin, 1981.

Food proteins (J. R. Whitaker and S. R. Tannenbaum eds). Westport: AVI Publishing Co, 1977.

R. J. Taylor, *The chemistry of proteins*. Unilever Educational Booklet, Advanced Series No 3, 1974.

R. A. Lawrie, *Meat science*, 3rd edn. Oxford: Pergamon Press, 1979.

J. C. Forrest *et al*, *Principles of meat science*. San Francisco: W. H. Freeman and Co, 1975.

Developments in meat science - 1 (R. Lawrie ed). London: Applied Science Publishers, 1980.

Breadmaking: the modern revolution (A. Williams ed). London: Hutchinson Benham, 1975.

Recent advances in the biochemistry of cereals (D. L. Laidman and R. G. Wyn Jones eds). Academic Press, 1979.

Developments in dairy chemistry - 1, Proteins (P. F. Fox ed). London: Applied Science Publishers, 1982.

Wheat chemistry and technology (Y. Pomeranz ed), 2nd edn. St Paul, Minnesota: American Association of Cereal Chemists, 1971.

5. Colours

Colour is very important in our appreciation of food. Mothers who may well deplore their children's selection of the most lurid items from the sweetshop will be found later in the greengrocery, applying identical criteria to their own selections of fruit. Since we can no longer expect to sample the taste of food before we buy it, appearance is really the only guide we have to quality other than past experience of the particular shop or manufacturer. Ever since food preservation and processing began to move from the domestic kitchen to the factory there has been a desire to maintain the colours of processed and preserved foods as close as possible to the colours of the original raw materials. Some foodstuffs acquire their recognised colours as an integral part of their processing. Examples are the brown of bread crust and other baked products considered in Chapter 2 and the pink of cured meats mentioned in Chapter 4. In this Chapter we will be largely concerned with the most colourful foodstuffs, fruit and vegetables, and the efforts of chemists to mimic their colours. Besides the three principle groups of plant pigments, the anthocyanins, the chlorophylls and the carotenoids the brown colours of polymerized, oxidised polyphenols and the purple-red betanins will be examined. The increasing use both of natural pigments such as cochineal and annatto and chemically synthesised dyes as food colorants also receive attention here.

The chlorophylls

The chlorophylls are the green pigments of leafy vegetables. They also give the green colour to the skin of apples and other fruit, particularly when it is unripe. Chlorophylls are the functional pigments of photosynthesis in all green plants. They occur, alongside a range of carotenoid pigments, in the membranes of the chloroplasts, the organelles which carry out photosynthesis in plant cells. The pigments of the chloroplast are intimately associated with other lipophilic components of the membranes such as phospholipids as well as the membrane proteins.

Algae and photosynthetic bacteria contain a number of different types of chlorophyll but the higher plants that concern us contain only chlorophylls a and b, in the approximate ratio of 3 to 1.

The structures of chlorophylls a and b are shown in *Fig. 5.1*. They are essentially porphyrins similar to those in haem pigments such as myoglobin except for the different ring substituents, the coordination of magnesium rather than iron, and the formation of a fifth ring by the linkage of position 6 to the γ methine bridge. By analogy with pigments such as myoglobin we can guess that the porphyrin ring will associate readily with the hydrophobic regions of chloroplast membrane proteins. The extended phytol side chain will similarly facilitate close association with carotenoids and membrane lipids.

Absorption spectra of chlorophylls a and b are shown in *Fig. 5.2*. The exact wavelengths of the absorption maxima, and their extinction coefficients, are not easily defined as they vary depending on the identity of the organic solvent used and the presence of traces of moisture. These variations are sufficient to make the accurate determination of chlorophyll levels in leafy tissues by spectrophotometry rather difficult. Another difficulty is that in solution, especially in solvents that include methanol, there is a tendency for oxidation or

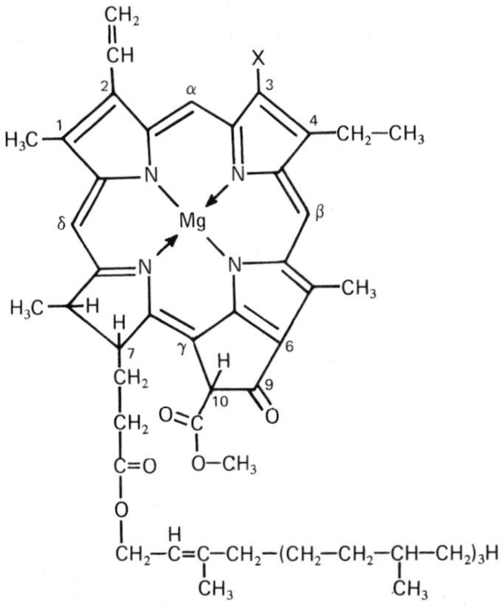

FIG. 5.1. Chlorophylls (X = $-CH_3$ in chlorophyll a; X = $-CHO$ in chlorophyll b). The double bonds of the porphyrin ring system are shown positioned quite arbitrarily. There is of course complete delocalisation of the π electrons in systems such as this. The isoprenoid alcohol esterified to the propionyl substituent at position 7 is known as *phytol*.

isomerisation to occur with replacement of the hydrogen at position 10 with $-OH$ or $-OCH_3$. It is chlorophyll's stability, or lack of it, in vegetable tissues, that is of interest to food chemists. Chlorophyll is lost naturally from leaves at the end of their active life on the plant. This breakdown accompanies a general breakdown of the choroplast membranes but the carotenoids are rather more stable so that autumn leaves and vegetables that are no longer fresh have a residual yellow colour. A number of different reactions have been implicated in chlorophyll breakdown including cleavage and opening of the porphyrin ring and removal of the phytol residue by the enzyme chlorophyllase, but there is very little known to support any of these hypotheses.

When green vegetables are heated, either in ordinary cooking, when they are blanched prior to freezing, or during canning there is evidence for the loss of the phytol side chain to give the corresponding chlorophyllide a or b but the most important event is the loss of the magnesium. This occurs most readily in acid conditions, the Mg^{2+} ion being replaced by protons, to give pheophytins a and b. Pheophytins have a dirty brown colour (see *Fig. 5.2*) and will be familiar as the dominant pigments in 'green' vegetables such as cabbage that have been overcooked. The acidity of the contents of plant cell vacuoles makes it difficult to avoid pheophytin formation especially during the rigorous heating involved in canning peas. One approach is to keep the cooking water slightly alkaline by the addition of a small quantity of sodium

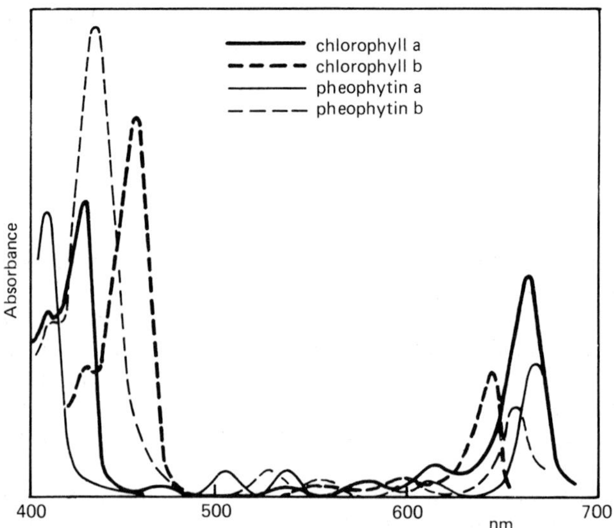

FIG. 5.2. Absorption spectra of chlorophylls and pheophytins.

bicarbonate. This is quite successful but the alkaline conditions have an unhappy effect on the texture and flavour and losses of vitamin C are enhanced.

In the canning of peas these colour losses have had to be accepted and artificial colours resorted to. It is said that cupric sulphate was occasionally, and dangerously, used to dye peas in the early years of this century but now organic chemical dyes are universally used, a mixture of Tartrazine and Green S being most popular. Growing doubt about the safety of such synthetic dyestuffs is encouraging food chemists to examine the possible use of chlorophyll derivatives as food colours. Chlorophyll itself is obviously unsuitable. Not only is it unstable but its insolubility in water makes it very difficult to apply. However, a derivative known as sodium copper chlorophyllin is now being used. This is the sodium salt of chlorophyllin (*ie* chlorophyll from which the phytol side has been removed) and which has the magnesium replaced by copper. It has an acceptable blue-green colour, is moderately soluble in water and, most important, does survive the heating conditions involved in canning. The amount of copper it contains is far too small in relation to the amount consumed to represent a toxicity hazard.

Carotenoids

Carotenoid pigments are responsible for most of the yellow and orange colours of fruit and vegetables. Chemically they are classed as terpenoids (see p. 144), substances derived in nature from the metabolic intermediate mevalonic acid [5.1], which provides the basic structural unit, the isoprene unit [5.2]. Terpenoids having one, two, three or four

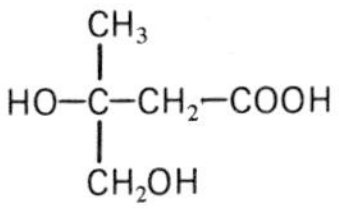

$$HO{-}C(CH_3)(CH_2OH){-}CH_2{-}COOH$$

5.1

$$-C-C{=}C-C- \text{ (with } -C- \text{ branch)}$$

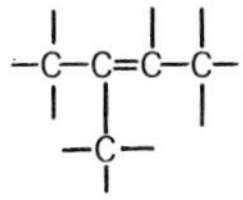

5.2

isoprenoid units (hemi-, mono-, sesqui- and diterpenoids respectively) are well known but to food chemists the steroids which are triterpenoids (*ie* 30 carbon atoms) and the carotenoids, the only known tetraterpenoid compounds, are the most important. Carotenoids occur in all photosynthetic tissues, alongside the chlorophylls, and in a number of non-photosynthetic plant tissues as components of chromoplasts. Chromoplasts are organelles which may be regarded as degenerate chloroplasts.

The carotenoids are divided into two principal groups, the carotenes, which are strictly hydrocarbons, and the xanthophylls which contain oxygen. The structures of most of the carotenoids important in food-

lycopene

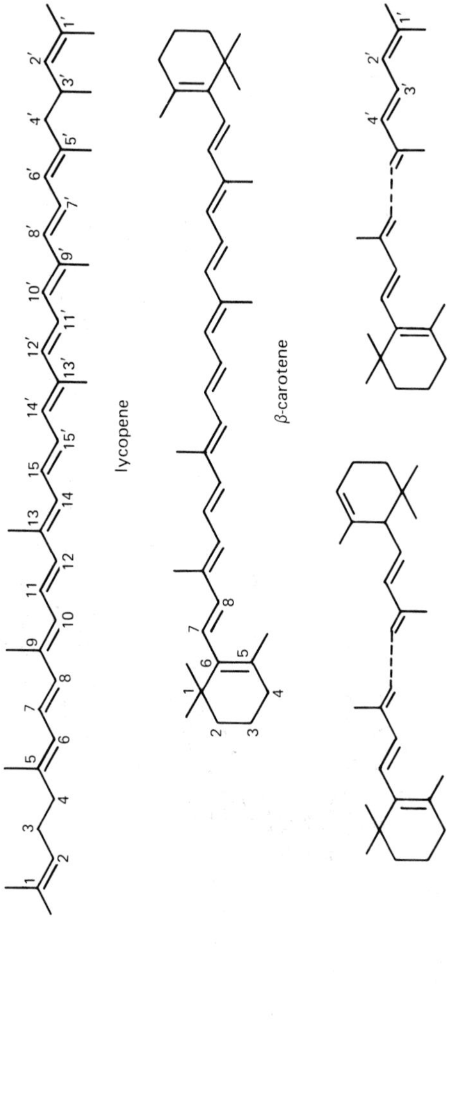

β-carotene

α-carotene

γ-carotene

FIG. 5.3. Carotene structures. The central region common to all carotenoids has been omitted here in α- and β-carotene and in *Fig. 5.4*.

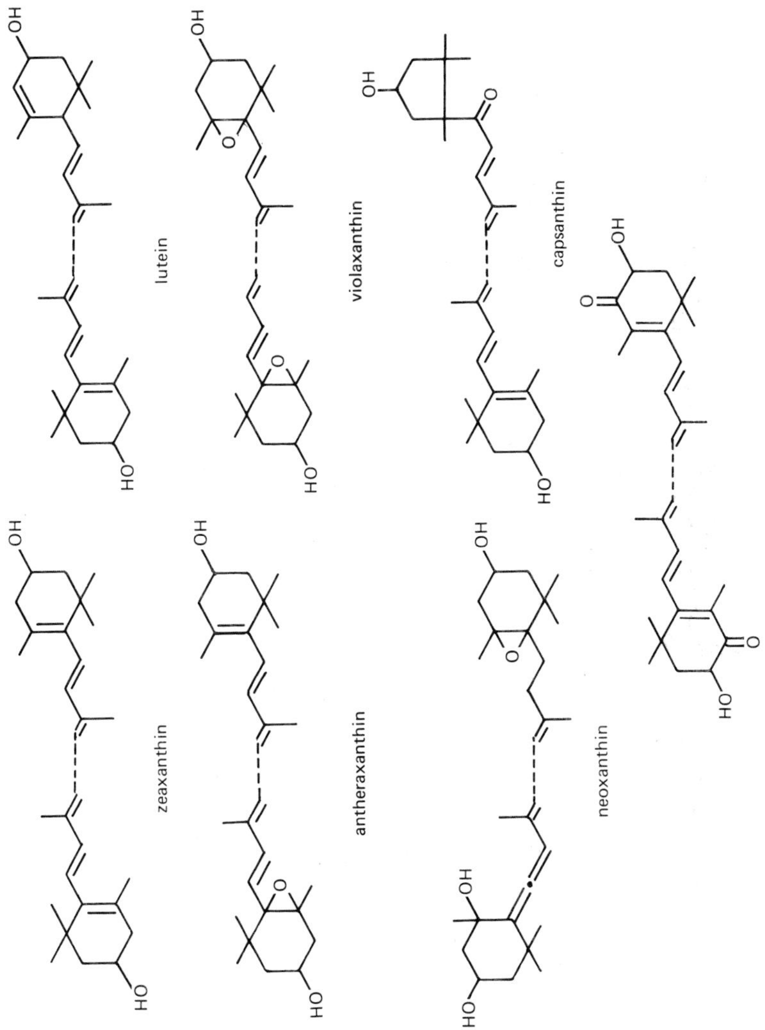

lutein

violaxanthin

capsanthin

astaxanthin

zeaxanthin

antheraxanthin

neoxanthin

FIG. 5.4. Xanthophyll structures.

stuffs are shown in *Figs 5.3* and *5.4*. The simplest carotene is lycopene and the numbering system shown illustrates clearly how the molecule is constructed from two diterpenoid subunits linked 'nose to nose'. Other carotenes are formed by cyclisation at the ends of the chain. Two possible ring structures result – the α-ionone [5.3] or the β-ionone [5.4].

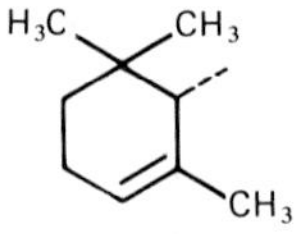

5.3

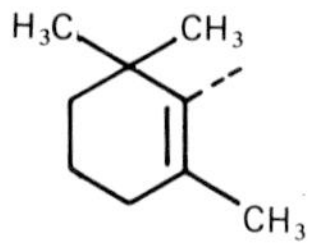

5.4

The acyclic end group occurring in lycopene and γ-carotene is sometimes referred to as the ψ-ionone structure.

The xanthophylls arise initially by hydroxylation of carotenes and most plant tissues contain traces of cryptoxanthins, the monohydroxyl precursors of the dihydroxyl xanthophylls such as zeaxanthin and lutein which are shown in *Fig. 5.4*. Subsequent oxidation reactions lead to the formation of epoxides such as antheraxanthin and violaxanthin and ketones such as astaxanthin and capsanthin. Neoxanthin is a rare natural example of an allene. The apocarotenoids are a small group of xanthophylls in which fragments have been lost from one or both ends of the chain. Three examples are shown in *Fig. 5.5*.

With the exception of the two acidic apocarotenoids, which will form water soluble salts under alkaline conditions, carotenoids are only freely soluble in non-polar organic solvents. Their absorbance spectra are generally rather similar, with the nature of the solvent causing nearly

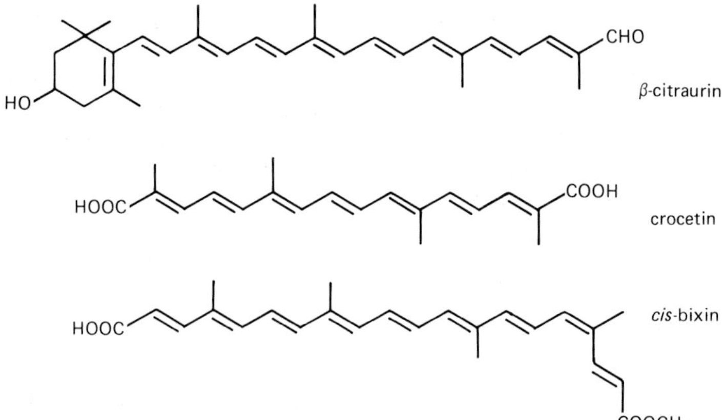

β-citraurin

crocetin

cis-bixin

FIG. 5.5. Apocarotenoid structures.

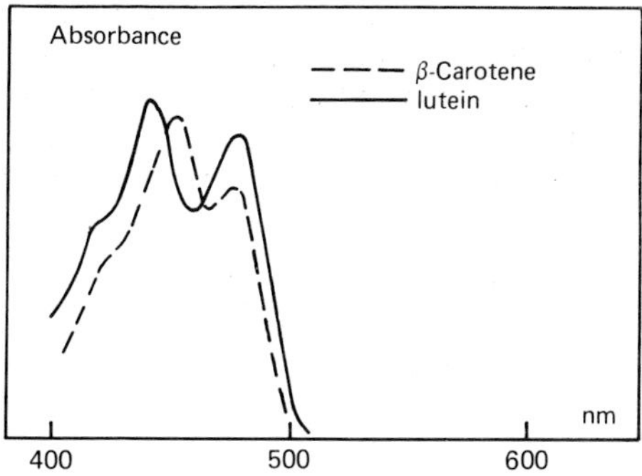

FIG. 5.6. Absorption spectra of carotenoids.

as much variation in λ_{max} and extinction coefficient values as is found between different carotenoids. The absorption spectra given in *Fig. 5.6* bear out the observation that xanthophylls are the dominant pigment in yellow tissues whereas carotenes tend to give an orange colour.

The distribution of carotenoids amongst the different plant groups or types of food materials shows no obvious pattern. Amongst the leafy green vegetables the carotenoid content follows the general pattern of all higher plant chloroplasts with β-carotene dominant and the xanthophylls lutein, violaxanthin and neoxanthin all prominent. Zeaxanthin, α-carotene, cryptoxanthin (3-hydroxy-β-carotene) and antheraxanthin also occur in small amounts. The quantity of β-carotene in leaf tissue is usually between 200 and 700μg per gram (dry weight). Amongst fruit a wider range of carotenoids is found. Only rarely (*eg* the mango and the persimmon) are β-carotene and its immediate xanthophyll derivatives cryptoxanthin and zeaxanthin predominant. In the tomato, lycopene is the major carotenoid. Orange juice contains varying proportions of cryptoxanthin, lutein, antheraxanthin and violaxanthin together with traces of their carotene precursors. The apocarotenoid β-citraurin has also been reported. During juice processing the acidic conditions promote some spontaneous conversion of the 5,6 and 5′,6′ epoxide groups to 5,8 and 5′,8′ furanoid oxides:

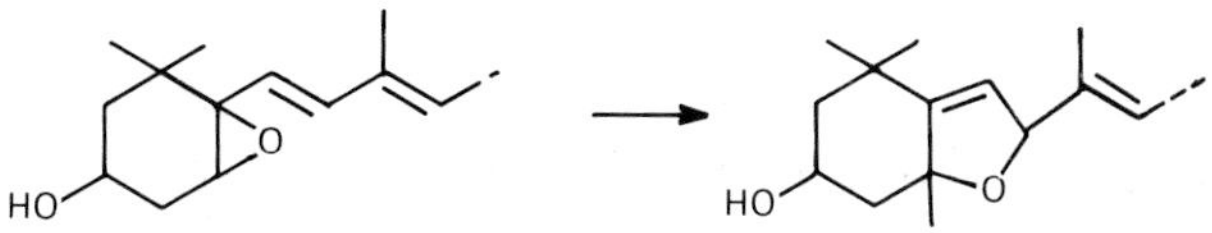

Some carotenoids are restricted to just a few or even a single plant species. The capsanthin of red peppers is a good example. Of course the classic occurrence of carotenoids is in the carrot. Here β-carotene predominates together with a proportion of α-carotene. The carotenoids of the usual varieties of carrot comprise only 5–10 per cent xanthophylls, located mostly in the yellow core. The total carotene content of carrots is 60–120 μg per gram fresh weight but some varieties are available with more than 300 μg per gram.

Partly due to their instability (see below) and partly their insolubility in water, carotenoids have, until recently, been little used as food colouring additives. Two exceptions are crocetin and bixin. Crocetin is the major pigment of the spice saffron, where it occurs in the form of a glycoside, each of its carboxyl groups being esterified to molecules of the disaccharide gentiobiose. Bixin is the major constituent of a colouring matter known as annatto. This is an extract of the fruit of the plant *Bixa orellana*, a large shrub grown in a number of tropical countries. Annatto is the traditional colouring for 'red' cheeses such as Leicester and is also used to colour margarine. Palm oil contains high levels of β-carotene (13–120 mg per 100 g) and is also used to colour margarine.

Three carotenoids have been chemically synthesised on a commercial scale, β-carotene, β-apo-8′-carotenal [5.5] and canthaxanthin [5.6].

These commercial preparations are being increasingly used in a wide range of products including margarine, cheese, ice-cream, and some baked goods such as cakes and biscuits.

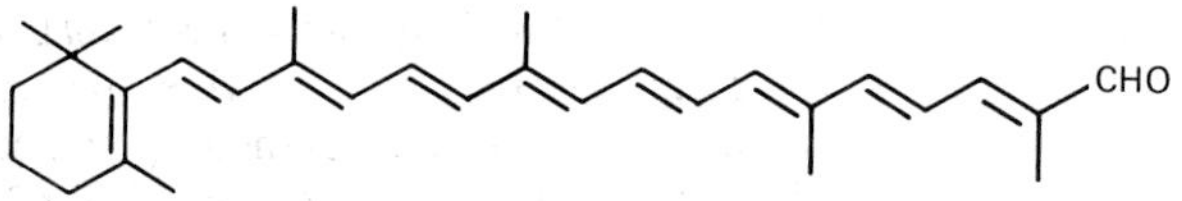

CHO

5.5

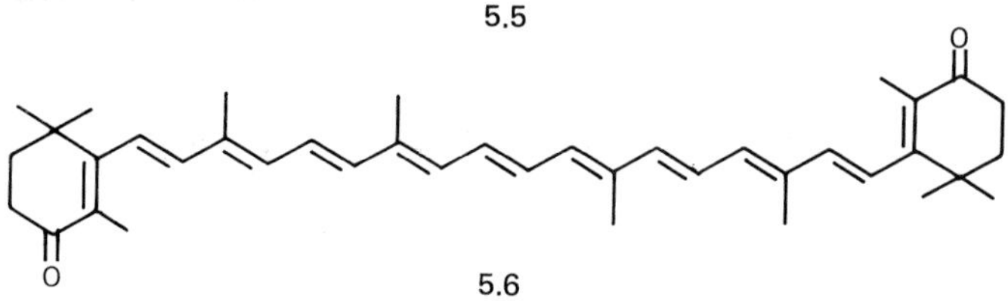

O

O

5.6

Although normally associated with plants, carotenoids do find their way into some animal tissues. Egg yolk owes its colour to the two xanthophylls, lutein and zeaxanthin, with only a small proportion of β-carotene. These carotenoids, and the smaller amounts that give the depot fat of animals its yellowish shade, are derived from the vegetable material in the diet. The dark greenish-purple pigment of lobster carapace is a complex of protein with astaxanthin. When lobster is boiled the protein is denatured and the colour reverts to the more typical reddish shade of a carotenoid. Astaxanthin is also the source of the pink colour of salmon flesh.

Carotenoids are generally quite stable in their natural environments but when food is heated or when they are extracted into solutions in oils or organic solvents they are much more labile. On heating in the absence of air there is a tendency for some of the *trans* double bonds of carotenes to isomerise to *cis*. The structures of the resulting 'neo-carotenes' have not been fully described but as the number of *cis* double bonds increases there is a loss of colour intensity. The difference in colour between canned and fresh pineapple provides a demonstration of the effect. The isomerisation of 5,6-epoxides to 5,8-furanoid oxides has already been mentioned. In the presence of oxygen, particularly in dried foods such as dehydrated diced carrots, oxidation and bleaching occurs rapidly. Hydroperoxides resulting from lipid breakdown are very effective in bleaching carotenoids. The final breakdown of β-carotene leaves a residue of β-ionone [5.7], a volatile compound which gives sun dried (and partially bleached) hay its characteristic odour.

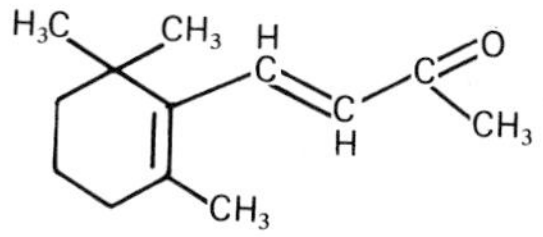

5.7

Although carotenoids as such have no physiological role in man the fact that some have vitamin A activity does give them nutritional significance. Vitamin A activity is possessed by substances that are precursors of retinal, the non-protein component of the visual pigment of the retina. The formation of retinal from carotenoid precursors containing the β-ionone ring, is shown in *Fig. 5.7*. Vitamin A is further discussed in Chapter 7.

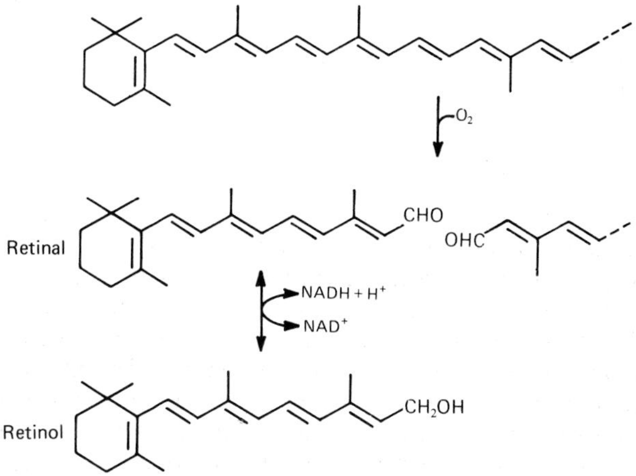

FIG. 5.7. Formation of retinol and retinal.

Anthocyanins

The pink, red, mauve, violet and blue colours of flowers, fruit and vegetables are caused by the presence of anthocyanins. In common with other polyphenolic substances anthocyanins occur in nature as glycosides, the aglycones being known as anthocyanidins. These are flavanoids, *ie* substances based on the flavan nucleus [5.8].

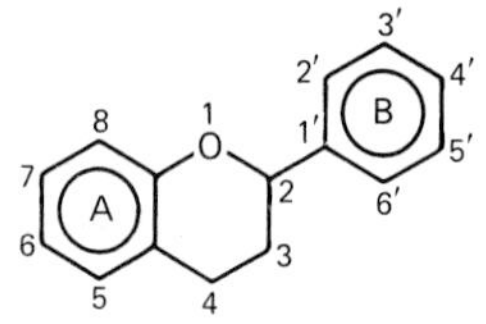

5.8

Six different anthocyanidins occur in nature but the diversity of the patterns of glycosylation means that there are innumerable different anthocyanins. A single plant species will also contain considerable numbers of different anthocyanins. The structures of the six anthocyanidins are shown in *Fig. 5.8*.

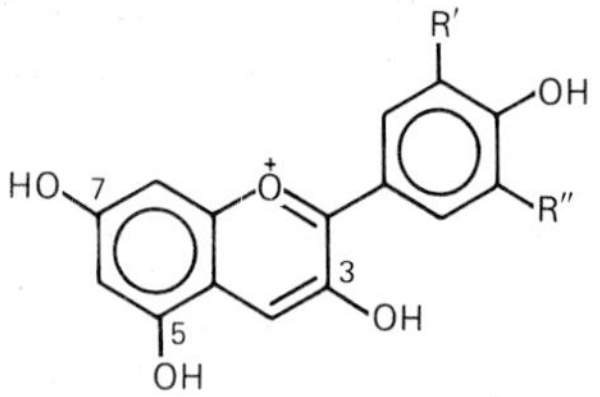

	R′	R″
pelargonidin	—H	—H
cyanidin	—OH	—H
peonidin	$—OCH_3$	—H
delphinidin	—OH	—OH
petunidin	$—OCH_3$	—OH
malvidin	$—OCH_3$	$—OCH_3$

FIG. 5.8. The anthocyanidins.

The form of the anthocyanidins shown in *Fig. 5.8* is correctly termed the flavylium cation, but as we shall see other forms occur depending on the pH. Anthocyanins always have a sugar residue at position 3 and glucose often occurs additionally at position 5 and more rarely at positions 7, 3′ and 4′. Besides glucose the monosaccharides most often found are galactose, rhamnose and arabinose. Some rather unusual disaccharides also occur such as rutinose (L-rhamnosyl α1-6 D-glucose) and sophorose (D-glucosyl β1-2 glucose). It is not uncommon for the 6 position of the sugar residue at position 3 to be esterified with phenolic compounds such as caffeic [5.9],

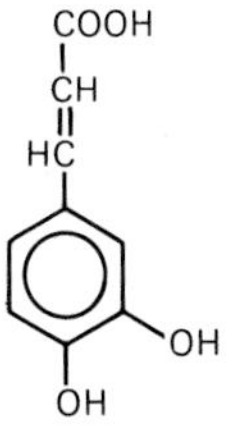

5.9 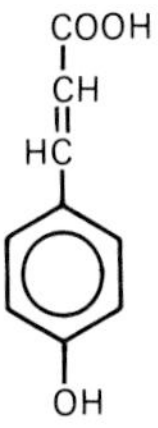 5.10 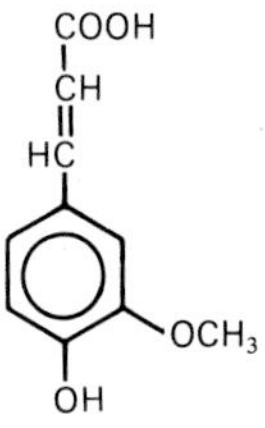 5.11

p-coumaric [5.10], or ferulic [5.11] acids. The principal anthocyanins of a range of fruit are listed in Table 5.1. Cyanidin is by far the most widespread anthocyanidin and it is clear that there is no link between the taxonomic classification of a plant and the identity of its anthocyanins.

Grape anthocyanins are particularly interesting. Of the six anthocyanidins only pelargonidin is not found in grapes and there is much more variety in the patterns of glycosylation and acetylation than in most plants. The classic European grape species is *Vitis vinifera* which contains only 3-monoglucosides whereas in the US and other non-European countries species such as *V. riparia* and *V. rupestris* and their hybrids with *V. vinifera* are grown. All these species, but not *V. vinifera*, contain both 3-monoglucosides and 3,5-diglucosides. The diglucosides can be readily detected by a chemical test and do also separate from the

Table 5.1. Fruit anthocyanins.

Fruit	Anthocyanidins	Glycosylation
Blackberry (*Rubus fructicosus*)	Cy.	3-gluc., 3-rutin.
Blackcurrant (*Ribes nigrum*)	Cy., Dp.	3-gluc., 3-rutin.
Raspberry (*Rubus ideaus*)	Cy., Pg.	3-gluc., 3-rutin, 3-soph.
Cherry (*Prunus* spp.)	Cy., Pn.	3-gluc., 3-rutin.
Strawberry (*Fragaria* spp.)	Pg., Cy.	3-gluc.
Grape (*Vitis vinifera*)	Mv., Dp., Pt., Pn., Cy.	3-gluc., some acetylation by p-coumaric and caffeic acids
Rhubarb (*Rheum* sp.)	Cy.	

Cy, cyanidin; Dp, delphinidin; Pg, pelargonidin; Pn, peonidin; Mv, malvidin; Pt, petunidin.

monoglucosides on chromatography. Detection of diglucosides in a red wine is therefore proof of its non-European origin. Much less work has been done on the anthocyanins of vegetables. Cyanidin is the anthocyanidin of red cabbage, pelargonidin occurs in radishes and red seeded varieties of beans and delphinidin occurs in aubergines.

Increasing substitution in the B ring moves the absorbance maximum of anthocyanidins towards the red end of the spectrum, *eg* λ_{max} for pelargonidin is 520 nm, for petunidin it is 543 nm, but these changes are not sufficient to account for the great range of colours that anthocyanins can give to flower petals and fruit. The pH and the presence of other substances have a much greater influence on their colour than the nature of the ring substituents. *Figure 5.8* shows the basic anthocyanidin structure in the flavylium cation form that predominates at low pH values; at pH 1 this is the only significant form. As the pH is raised a proton is lost, a water molecule is acquired and the carbinol pseudo-

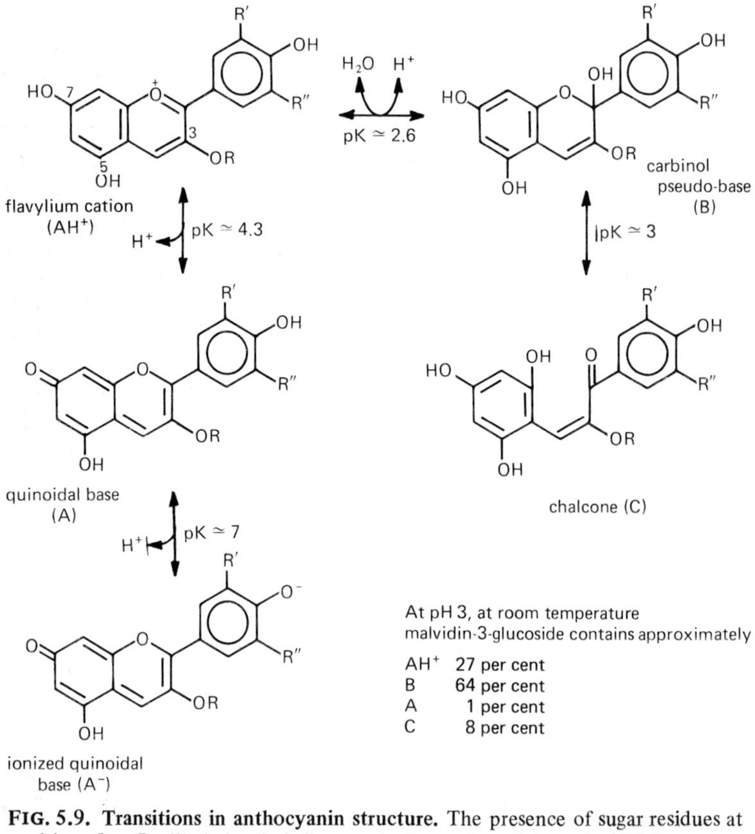
flavylium cation (AH⁺)
H₂O H⁺
pK ≃ 2.6
carbinol pseudo-base (B)
pK ≃ 4.3
pK ≃ 3
quinoidal base (A)
chalcone (C)
pK ≃ 7
ionized quinoidal base (A⁻)
At pH 3, at room temperature malvidin-3-glucoside contains approximately
AH⁺ 27 per cent
B 64 per cent
A 1 per cent
C 8 per cent

FIG. 5.9. Transitions in anthocyanin structure. The presence of sugar residues at positions 5 or 7 will obviously influence the formation of the quinoidal forms.

base is formed (see *Fig. 5.9*). The pK value for this transition is around 2.6. The flavylium cations are the red-coloured forms of the anthocyanins, the carbinol pseudo-bases are colourless, so that there is a gradual loss of colour intensity with rise in pH. Only recently has work with purified anthocyanins begun to clarify the pattern of the other ionizations including those at higher pH values. Above pH 3 the small residual proportion of the anthocyanins in the flavylium cation form begin to lose protons, forming first the quinoidal base, which is weakly purple in colour, and then the ionized quinoidal base, which is deep blue. At the same time the carbinol pseudo-base gives rise to a small proportion of the colourless chalcone forms. After extended periods at elevated pH quite high proportions of chalcones are formed. Although these equilibria have been described in some detail for purified anthocyanins it must be remembered that in fruit juices and other natural plant materials the situation is neither so straightforward nor so well understood. Anthocyanins readily interact with other colourless flavanoids which abound in plant tissues. These associations, which presumably depend on hydrogen bonding between hydroxyl groups, enhance the intensity of the colour, modify λ_{max} values, and tend to stabilise the blue quinoidal forms.

Anthocyanins also complex metal cations but it is not known how important this is to the colour of fruit. Occasional unusual colour developments in canned fruit are undoubtedly due to interactions between the metal of the can (when the internal lacquering has failed) and the anthocyanins of the fruit. Another important reaction of anthocyanins is that with sulphur dioxide. Sulphur dioxide, usually as sulphite (SO_3^{2-}) or metabisulphite ($S_2O_5^{2-}$) (see Chapter 8) is routinely used as an antimicrobial preservative in wines and fruit juices. At high concentrations (1–1½ per cent) it causes a total irreversible bleaching* of anthocyanins but at lower concentrations (500–2000 ppm) it reacts with the flavylium cation to form a colourless addition compound – the chroman-2-sulphonic acid [5.12]. Acidification or the addition of excess acetaldehyde (ethanal) will remove the SO_2.

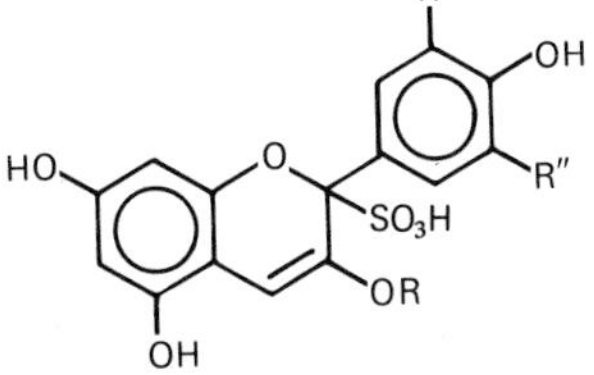

R′ OH R″ HO O SO_3H OR OH

5.12

*Maraschino cherries are bleached with SO_2 prior to dyeing them to appropriate colours to garnish cocktails and similar drinks.

As a red wine matures the anthocyanins slowly undergo a number of reactions with the other colourless flavanoids that were also originally present in the grape skin such as derivatives of catechin [5.13], a flavanol.

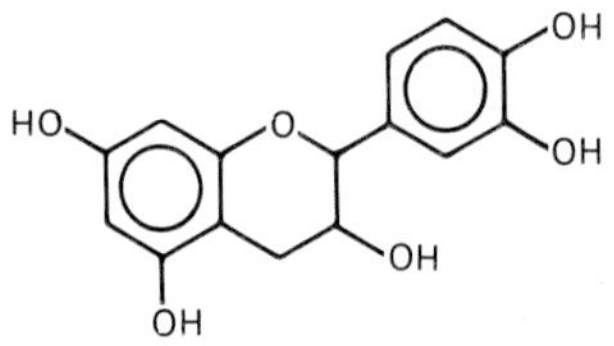

5.13

The formation of a link between the 4 position of the anthocyanin and the 8 position of the catechin leads to compounds whose colour is enhanced and less affected by pH change or SO_2. The brown of old wines is the result of extensive polymerization of the anthocyanins and other flavanoids. This is accompanied by a mellowing of the flavour as it is these other flavanoids, usually described as tannins, which give red wine its characteristic astringency.

In most food processing operations the anthocyanins are quite stable, especially when the low pH of the fruit is maintained. Occasionally, however, the ascorbic acid naturally present can cause problems. In the presence of iron or copper ions and oxygen the oxidation of ascorbic acid to dehydroascorbic acid is accompanied by hydrogen peroxide formation. This will oxidise anthocyanins to colourless malvones [5.14], a reaction implicated in the loss of colour by canned strawberries.

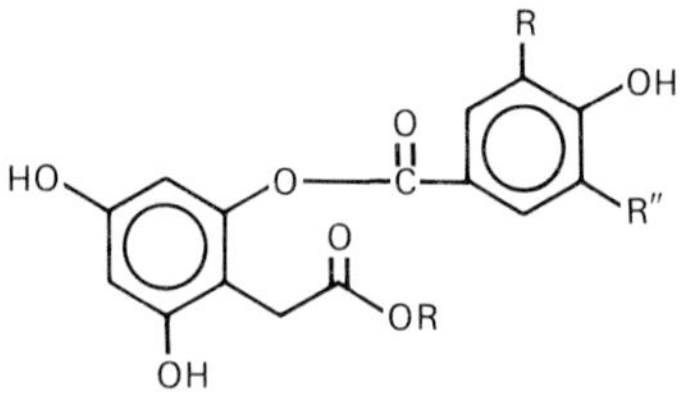

5.14

Betalaines

Until quite recently the characteristic red-purple pigments of beetroot, *Beta vulgaris* were described as 'nitrogenous anthocyanins'. However, it is now established that these pigments, and those of the other members of the eight families usually grouped together as the order *Centrospermae* are quite distinct from the anthocyanins and never occur with anthocyanins in any member of these eight families. The betalaines are divided into two groups, the betacyanines, which are purplish red in colour (λ_{max} 534-555 nm), and the less common, yellow, betaxanthines ($\lambda_{max} \sim 480$ nm). The betalaines of *B. vulgaris* are the only ones of food significance and this account will therefore be restricted to them. Their structures are shown in *Fig. 5.10*.

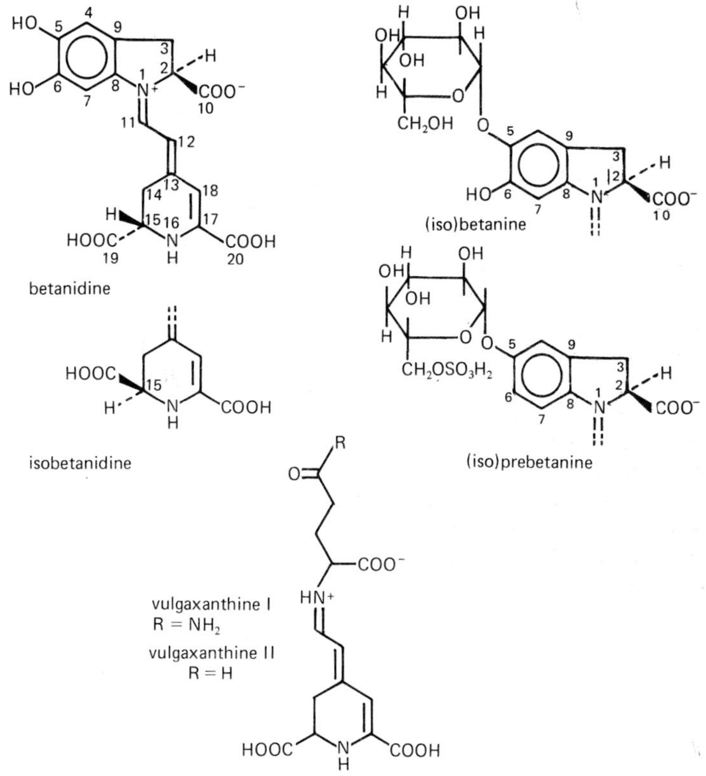

FIG. 5.10. Betalaine structures.

Firstly, we have the betacyanines, which comprise about 90 per cent of beetroot betalaines. They are all either the glycosides or the free aglycones of betanidine or its C-15 isomer. In beetroot, iso forms account for only 5 per cent of the total betacyanines. About 95 per cent of the betanidine and isobetanidine carry a glucose residue at C-5, and a very small proportion of these glucose residues are ésterified with sulphate. Secondly, we have the betaxanthines, in beetroot represented by the vulgaxanthines, characterised by the lack of an aromatic ring system attached to N-1, and the absence of sugar residues. A great diversity of betaxanthines are found in the *Centrospermae* but only the two known as vulgaxanthines I and II are found in beetroot, in roughly equal proportions.

Although of a similar colour to the other water soluble plant pigments, the anthocyanins, betacyanines differ in that their colour is hardly affected by pH changes in the range normally encountered in foodstuffs. Below pH 3.5 the absorption maximum of betanine solutions is 535 nm, between pH 3.5 and 7.0 it is 538 nm and at pH 9.0 rises to 544 nm. Betacyanins are fairly stable under food processing conditions, although heating in the presence of air at neutral pH values causes breakdown to brown compounds. They have therefore been examined as possible food colorants for confectionery. For this application the beetroot juice is first fermented with yeast to eliminate its high sugar content and then dried to give a powder having 6–7 per cent of betacyanin.

Melanins

Although we do not necessarily regard them as desirable pigments in our fruits and vegetables this is a convenient point to consider the brown, melanin-type pigments that arise when plant tissues are damaged. Everyone is familiar with the way in which pale coloured fruit and vegetables such as apples, bananas or potatoes quickly turn brown if air is allowed access to a cut surface. The browning occurs when polyphenolic substances, which are usually contained within the vacuole of the plant cells, are oxidised by the action of the enzyme phenolase, which occurs in the cytoplasm of plant cells. Tissue damage caused by slicing or peeling, fungal attack or bruising, will bring enzyme and substrates together. The enzyme phenolase (originally known as polyphenol oxidase) occurs in most plant tissues and a related enzyme, tyrosinase, is found in mammalian skin. Phenolase is an unusual (but not unique) enzyme in that it catalyses two quite different types of reaction. The first of these, its cresolase activity, results in the oxidation of a monophenol to an *ortho*-diphenol:

OH, R; $\frac{1}{2}O_2$ cresolase activity → OH, OH, R; $\frac{1}{2}O_2$, H_2O catecholase activity → O, O, R

The second, catecholase, activity oxidises the *o*-diphenol to an *o*-quinone. *O*-quinones are highly reactive and readily polymerize after their spontaneous conversion to hydroxyquinones:

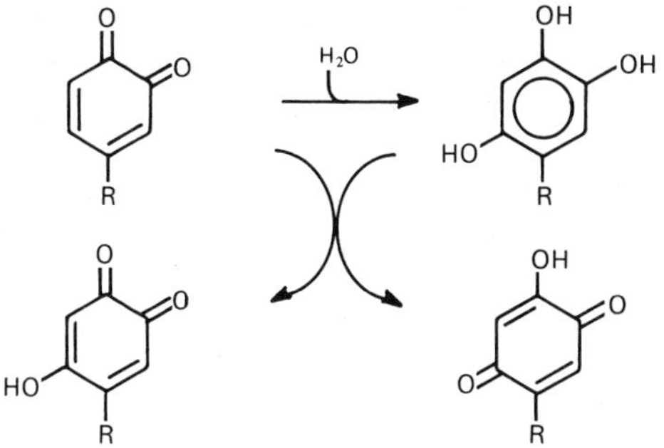

Interactions like these with the tyrosine residues of enzyme proteins will lead to enzyme inactivation. In animal tissues the only substrate is the amino acid tyrosine. The reaction in this case leads to the melanin pigments of skin and hair. The brown pigments that develop on the cut surface of fruit and vegetables may well be regarded as unsightly to consumers but they do not detract from the flavour or nutritional value. To the plant the products of phenolase action, particularly the quinones, are very important as antifungal agents, reducing the ability of fungi to penetrate tissues that have suffered minor physical damage. The relative resistance to fungal diseases of different varieties of onions and apples has been correlated with their phenolase activities and quinone contents.

The most important substrate for phenolase in apples, pears and potatoes is chlorogenic acid [5.15], and in onions protocatechuic acid [5.16] but very little is known about the structures of the polymerized

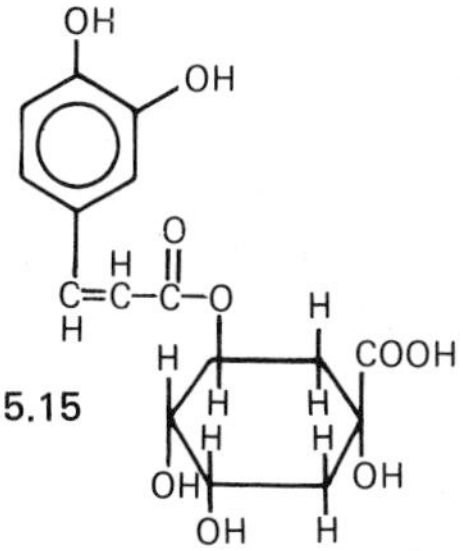

5.15

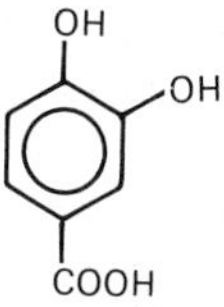

5.16

quinones that are derived from them.

During fruit and vegetable processing (in the home as well as the factory) we try to prevent phenolase action. Making sure that the fruit and vegetables are blanched as soon as possible after (or even before) any tissue damaging operations will reduce phenolase action to a minimum. Reducing contact with air by immersion in water is also common practice. Some of the most effective specific inhibitors of phenolase activity are chelating agents that act by depriving the enzyme of the copper ion that is a component of active site. However, agents such as diethylthiocarbamate, CN^-, H_2S or 8-hydroxyquinoline are hardly suitable for food use.

Though less specific and less potent the organic acids citric acid [5.17] and malic acid [5.18] do have advantages. Not only are they

$$\begin{array}{c} H_2C\text{—}COOH \\ | \\ HO\text{—}C\text{—}COOH \\ | \\ H_2C\text{—}COOH \end{array}$$

5.17

$$\begin{array}{c} COOH \\ | \\ H\text{—}C\text{—}OH \\ | \\ CH_2 \\ | \\ COOH \end{array}$$

5.18

common components of fruit juices but their acidity also helps to keep the pH well below the optimum for phenolase action (approx. pH 7). Immediately after peeling or slicing, fruit such as peaches are often immersed in baths containing dilute solutions of these acids, frequently supplemented with ascorbic acid or sulphite.

Ascorbic acid is not only valued as a vitamin but it is also a good chelating agent and antioxidant. It interacts directly with the quinones to prevent browning and if sufficient is used will 'mop up' the oxygen in closed containers such as cans:

$$o\text{-diphenol} + \tfrac{1}{2}O_2 \rightarrow o\text{-quinone} + H_2O$$

$$o\text{-quinone} + \text{ascorbate} \rightarrow o\text{-diphenol} + \text{dehydroascorbate}$$

It seems likely that the manufacturers of dehydrated potato powders include ascorbic acid in their products to ensure its whiteness rather than its vitamin C activity.

Sulphur dioxide also has a specific, if not fully explained, inhibitory action against phenolase. A sulphite concentration sufficient to maintain a free SO_2 concentration of 10 ppm will completely inhibit phenolase. Being a reducing agent sulphite has the additional benefit of preserving the ascorbic acid level but it is not without its disadvantages as it bleaches anthocyanins and also hastens can corrosion.

The enzymic oxidation of polyphenolic substances is a highly desirable feature of one particular foodstuff – tea. The tea we drink is a hot water infusion of the leafy shoots of the tea plant, *Camellia sinensis*, which has been processed in various ways depending on the type of tea to be produced. The type of tea that most of us drink is known as 'black tea' to distinguish it from the 'green tea' we usually call 'China tea'. The leaves of the tea plant contain enormous quantities of polyphenolic materials – over 30 per cent of the dry weight. Caffeine [5.19], the stimulant found also in coffee and 'cola'-based drinks is

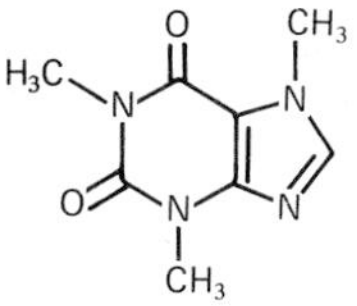

5.19

some 3–4 per cent of the dry weight of the leaf. After the freshly plucked leaves have been withered, *ie* allowed to lose about one fifth of their water content, the leaves are macerated so that the phenolase of the leaf can come into contact with the polyphenolic substances of the cell vacuoles. The macerated leaves are then left at ambient temperatures for a few hours to 'ferment' during which time a high proportion of the polyphenolic substances are oxidised and much polymerization occurs. The leaves are then 'fired', *ie* dried out at temperatures up to 75 °C, to bring the reactions to a halt and give us the material we know as tea.

Almost all the polyphenolic substances in tea are flavanols – either catechin itself or its derivatives as shown in *Fig. 5.11*. *Epi*-gallocatechin

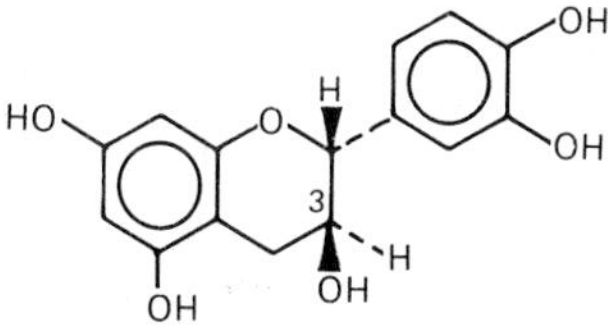

catechin

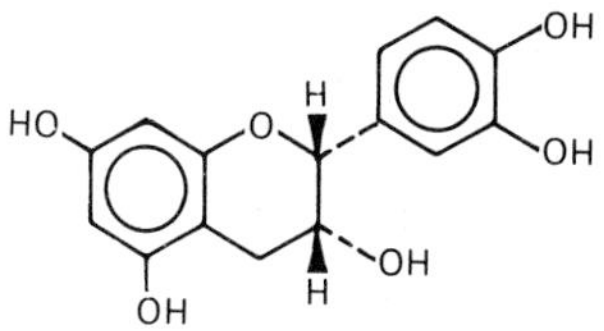

epi-catechin

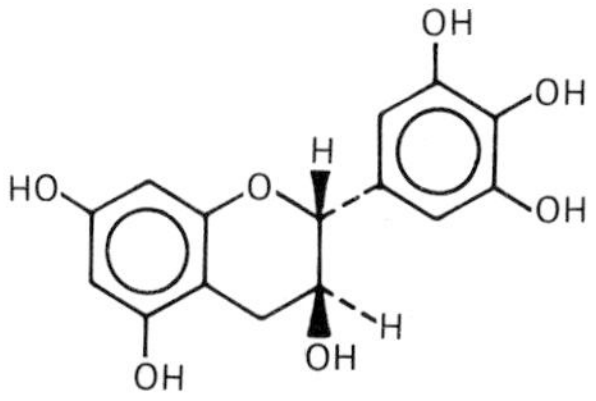

gallocatechin

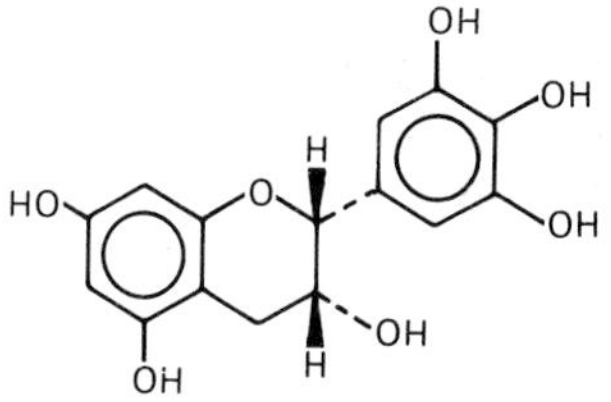

epi-gallocatechin

Note:
Epi-catechin and epi-gallocatechin also occur as gallates, with gallic acid:

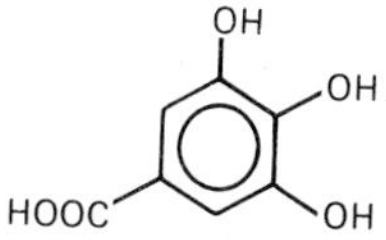

esterified at the 3-position of the flavanol.

FIG. 5.11. The principal flavanols of fresh tea flush. The tea flush is the unfermented shoot consisting of the bud and the first and second leaves.

gallate is by far the most abundant of the flavanols. The initial effect of phenolase action is in every case to convert the flavanol to the corresponding *o*-quinone:

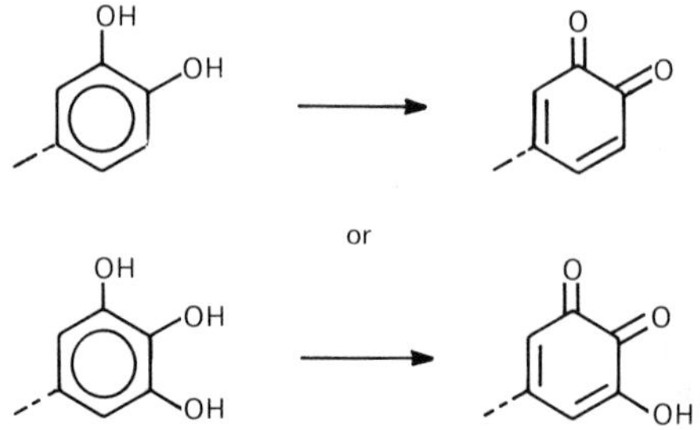

or

The quinones are highly reactive and readily form a wide variety of dimers. Most of these interact further to produce the polymeric thearubigins but one class of dimers, the theaflavins [5.20], comprise about 4 per cent of the soluble solids of black tea.

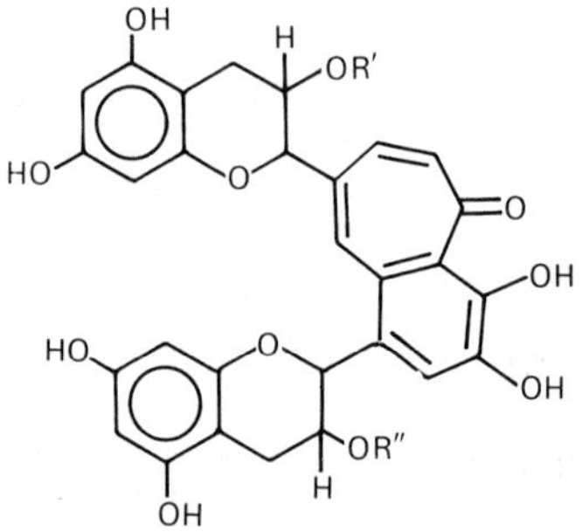

R′ and R″ may be either —H or gallate residues

5.20

Being bright red in colour they make a major contribution to the 'brightness' of tea colour. The brown thearubigins are the principal pigments of black tea infusions. They contribute not only the bulk of the colour but also the astringency, acidity and body. Thearubigins that are extracted with hot water have a wide range of molecular weights. Although their average molecular weight corresponds to oligomers of four or five flavanoid units molecules of up to some 100 flavanoid units are found in tea infusions. Degradation studies have shown that all the principal flavanols shown in *Fig. 5.11* are represented. Although there is no certainty as to the nature of the links involved in the polymerization a proanthocyanidin structure is most likely. The proanthocyanidins themselves are dimers which occur in small amounts in many fruit juices where they contribute to the astringency. As shown here the formation of the link is a reductive reaction which in the tea fermentation would almost certainly be coupled to other oxidative processes:

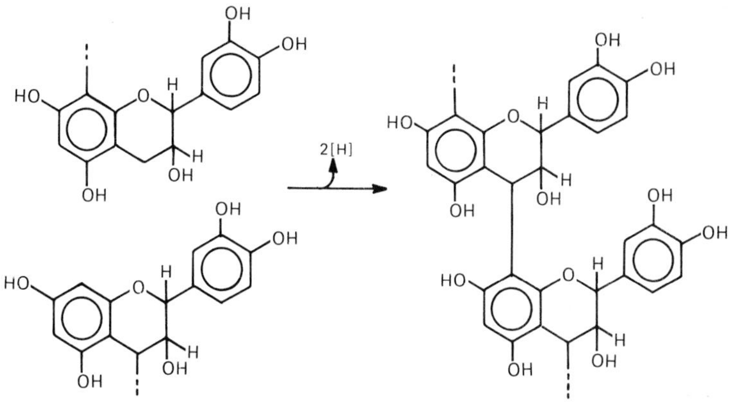

Most of the substances that we have considered so far in this chapter have been important for the colours they impart to the food materials where they occur naturally. Some natural pigments such as the betacyanines and β-carotene are now being extracted from natural sources or chemically synthesised for use as colorants in foods that would not otherwise contain them. In the cases of bixin and crocetin we have pigments that occur in plant materials whose only value as food ingredients is as a source of colour. Plants do not, however, have a monopoly in the supply of natural colouring materials. Cochineal is the name given to a group of red pigments from various insects, similar to aphids, which belong to the super-family *Coccidoidae*. Cochineal, in the form of the dried and powdered female insects (about 100 000 per kilo) has been an object of commerce since ancient times. Besides its use as a food colour it has been used in a dyestuff for textiles and leather and also a heart stimulant. Although varieties of cochineal have been 'cultivated' in many parts of the world the bulk of world production now comes from Peru where it is obtained from insects that are parasites on cacti.

The principle component of cochineal is carminic acid [5.21] an anthroquinone glycoside with an unusual link between the glucose and the aglycone.

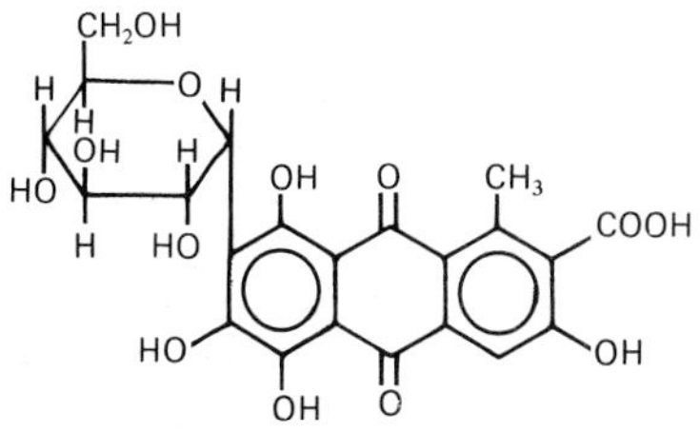

5.21

Solutions of carminic acid itself are uncoloured. The deep red coloured 'carmines of cochineal' are obtained by first extracting the carminic acid from the crude cochineal powder with hot water. The extracts are then treated with aluminium salts to produce brilliantly coloured complexes of uncertain structure. These can be precipitated out with ethanol to give a water soluble powder. In spite of cochineal's high cost the increasing unpopularity of chemically synthesised dyestuffs is bound to lead to a revival in use.

Artificial food colorants

The use of artificial, unnatural colours in food has a long and far from glorious history. Wine was always vulnerable to the unscrupulous and in the 18th and 19th centuries burnt sugar and a range of rather unpleasant 'vegetable' extracts were commonly used to give young red wines the appearance of mature claret. However, for sheer lethality the confectionery trade had no equal. In 1857 a survey of adulterants to be found in food revealed that sweets were commonly coloured by, for example, lead chromate, mercuric sulphide, lead oxide (red lead) and copper arsenite! Legislation and the availability of the newly developed aniline dyestuffs, rapidly eliminated the use of metallic compounds as well as some of the more dubious vegetable extracts. The new chemically synthesised dyestuffs had many advantages over 'natural' colours. They were much brighter, more stable, cheaper, and offered a wide range of shades. It did not take long before the toxic properties of these dyes also became apparent, though mostly through their effects on those engaged in making them rather than on consumers.* Since that time there has been a steady increase in the range of dyestuffs available but at the same time we have become increasingly aware of their toxicity. Thus in 1957 there were 32 synthetic dyestuffs permitted for use in food in Britain. By 1973 19 dyestuffs had been removed from the list and three added. In 1979 the list was shortened to 11 as British legislation came into line with that of the EEC. Rather as one might expect most countries outside the EEC enforce their own views of what should or should not be regarded as safe. Norway and Sweden have taken the extreme approach of banning all synthetic dyestuffs whereas the US has a list very similar to our own, with just sufficient differences to cause occasional problems for exporters of confectionery and other food products.

The structures of the 11 synthetic colours permitted in the UK are shown in *Fig. 5.12*.

*It is said that the laxative effects of phenolphthalein were only discovered when wine, to which it had been added to enhance the colour, had quite unanticipated effects on the drinkers.

Azo

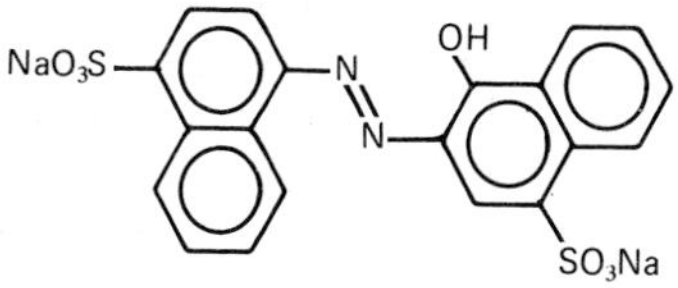

Carmoisine

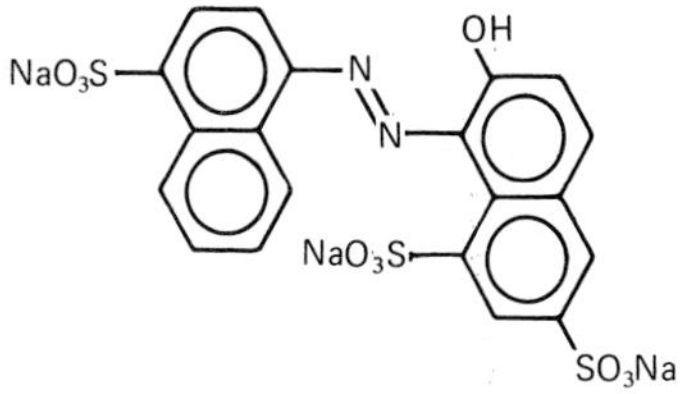

Ponceau 4R

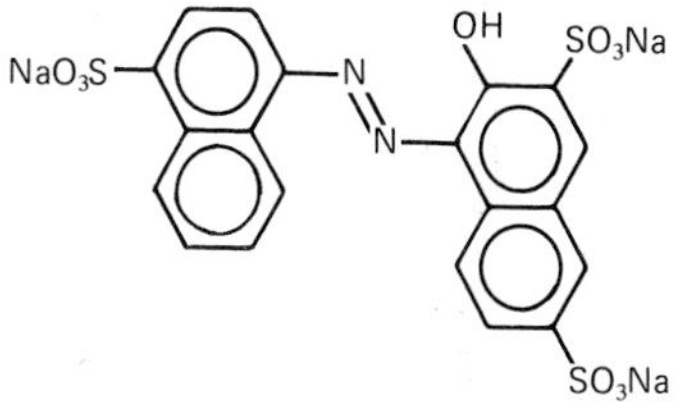

Amaranth

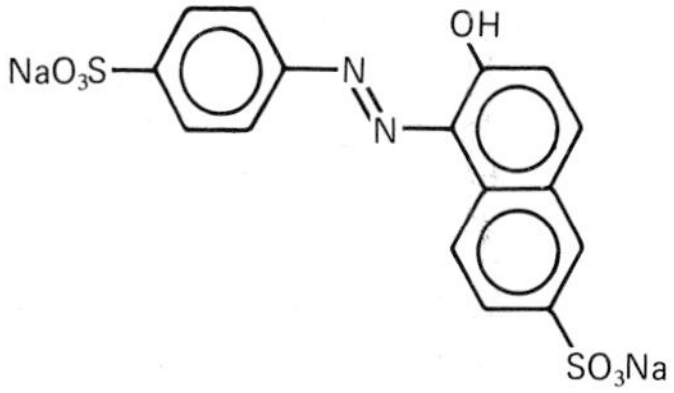

Sunset Yellow FCF

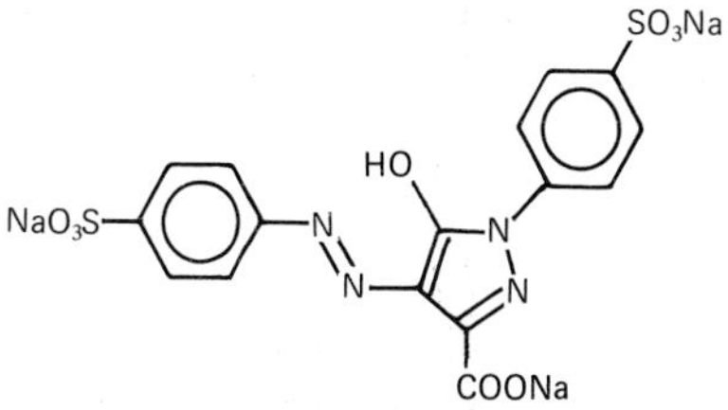

Tartrazine

FIG. 5.12. The synthetic food colours permitted in the UK – classified by basic chemical structure.

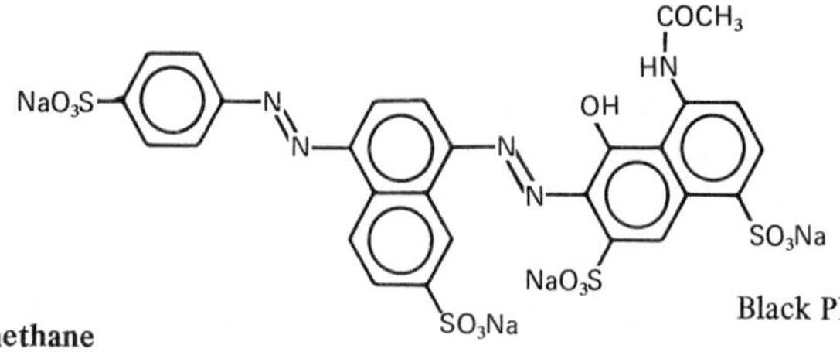

Black PN

Triarylmethane

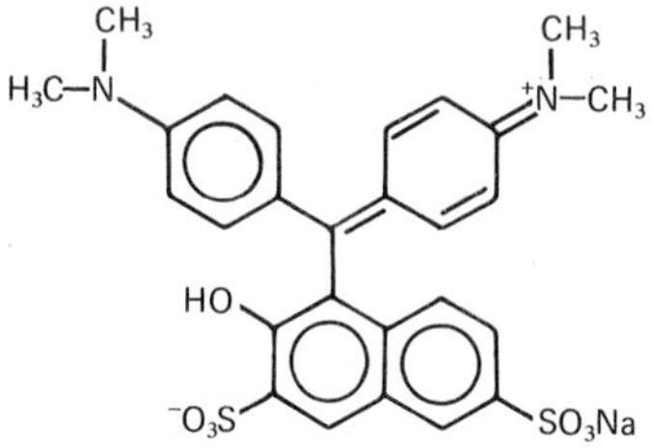

Food Green S

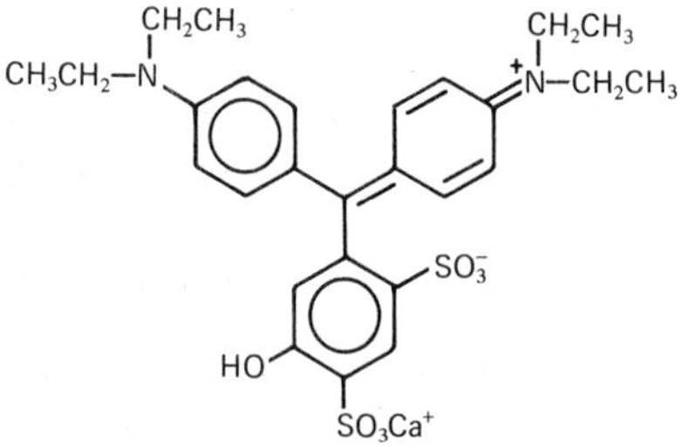

Patent Blue V

Xanthene

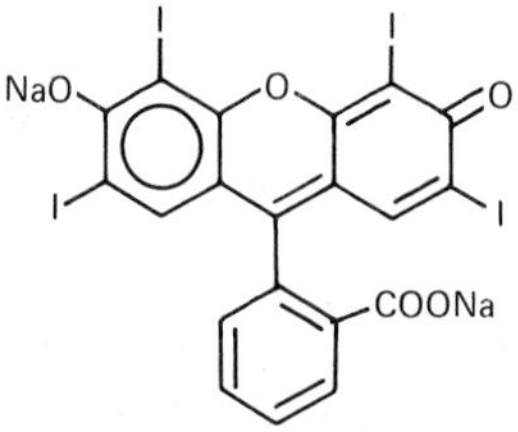

Erythrosine

Quinoline

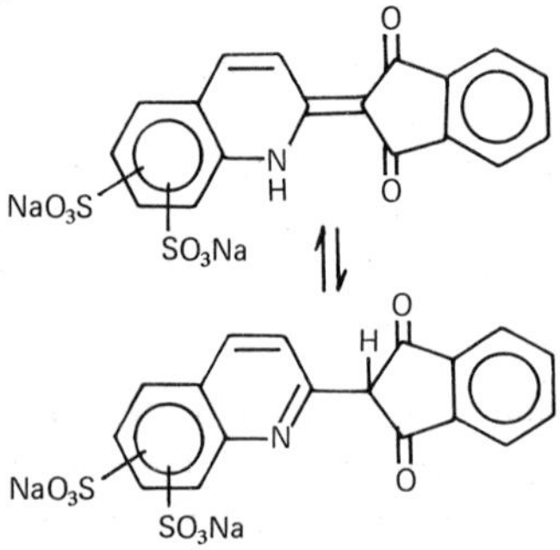

Quinoline Yellow

Indigoid

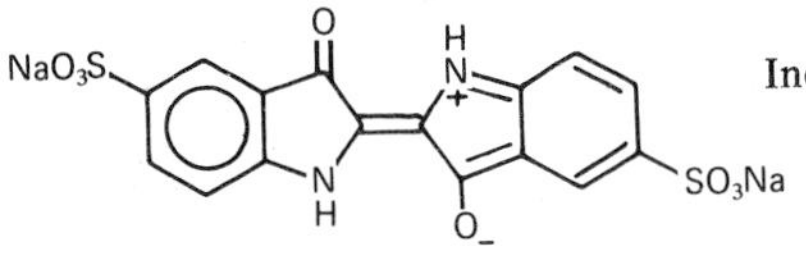

Indigo carmine

Four other colours, Red 2G, Yellow 2G, Brown FK and Chocolate Brown HT, all azo compounds, are provisionally permitted for the time being but are being subjected to special scrutiny. Examination of these structural formulae shows that water solubility is mostly conferred by the presence of sulphonic acid groups. The chromophoric group of the azo dyes is the azo group in association with one or more aromatic systems giving mostly colours in the yellow, orange, red and brown range. The three linked aromatic systems provide the chromophoric group of the triarylmethane compounds which are characteristically bright green or blue. The chromophoric group of xanthene compounds is similar to that of the triarylmethane compounds and imparts a brilliant red shade to Erythrosine, the only permitted example. Resonance hybrids are important in the colours of both quinoline and indigoid compounds. Indigo carmine in particular has many possible zwitterion arrangements besides that shown in *Fig. 5.12*. The absorption spectra of these dyes are shown in *Fig. 5.13*.

The application of the synthetic food colours in food processing demands attention to many other factors besides the shade and intensity required. For example in sugar confectionery and baked goods high cooking temperatures may pose problems. In soft drinks one may have to contend with light (Food Green S, Erythrosine and Indigo Carmine are particularly vulnerable), sulphur dioxide, low pH and ascorbic acid. Probably the greatest use of food colours is in fruit-based products. Table 5.2 lists the dyestuffs used to supplement or mimic the natural fruit colour in a range of products. Canned peas are coloured with a mixture of Food Green S and Tartrazine. Brown colours in chocolate and caramel products are obtained either with Chocolate Brown HT or mixtures such as Sunset Yellow and Amaranth, Food Green S and Tartrazine. Canned meats can be coloured with Erythrosine but products such as sausages, or meat pastes, which will be exposed to light are usually coloured with Red 2G.

Current regulations in Britain forbid the addition of any colour to raw or unprocessed foods such as meat, fish and poultry, fruit and vegetables and tea and coffee (including 'instant' products). Although not expressly forbidden at the present time the use of added colours in baby foods is strongly discouraged.

In general, food colours are added at a rate of 20–100 mg per kg. It has been calculated that in the UK there is an average consumption of about 0.5 kg of coloured food per day which happily corresponds to daily consumption figures of the various colours which are in all cases less than 10 per cent, and usually less than 1 per cent of the 'Acceptable Daily Intake' figures that are now being established by toxicologists.

The ever-increasing costs of safety testing of new food additives is making the appearance of new food colours extremely unlikely. It must be remembered that it is not only the food colour itself that must be shown not to cause death or disease but also any impurities that arise during its manufacture, and breakdown products that might arise during food processing operations, cooking or digestion have to be cleared.

Table 5.2. Food colours for use in fruit products.

Product	'Flavour'				
	Strawberry	Raspberry	Blackcurrant	Orange	Lemon
Soft drinks	Cm + SY	Cm + Pc	Am	SY	Tz
Sweets and ice cream	Pc	Cm	Am	SY	Tz
Jam	Pc	Cm	Rd	SY	Tz
Canned fruit	Pc + SY	Cm	Am	–	–

Key: Cm, carmoisine; Pc, Ponceau 4R; SY, Sunset Yellow; Am, Amaranth; Rd, Red 2G; Tz, Tartrazine.

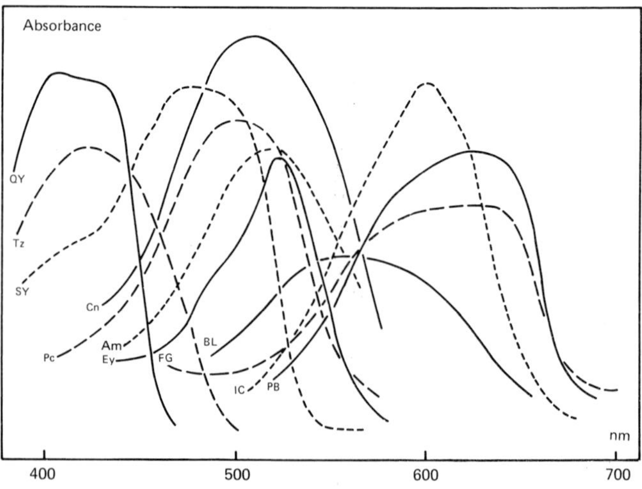

FIG. 5.13. The absorption spectra of aqueous solutions of the common food colours (at pH 4.2 in citrate buffer). (Key: As for Table 5.2 plus IC, Indigo Carmine: FG, Food Green S; Ey, Erythrosine; QY, Quinoline Yellow; BL, Black PN; PB, Patent Blue V.

Further reading

T. W. Goodwin and E. I. Mercer, *Introduction to plant biochemistry*. Oxford: Pergamon Press, 1983.

Natural colours for food and other uses (J. N. Counsell ed). London: Applied Science Publishers, 1981.

Development in food colours - 1 (J. Walford ed). London: Applied Science Publishers, 1980.

R. J. Taylor, *Food additives*. Chichester: John Wiley & Sons, 1980.

J. B. Harborne, *Phytochemical methods, a guide to modern techniques of plant analysis*. London: Chapman and Hall, 1973.

Chemistry and biochemistry of plant pigments (T. W. Goodwin ed). London: Academic Press, 1976.

6. Flavour

However much nutritionists and 'health food' enthusiasts may wish otherwise, it is the flavour and appearance of food rather than the vitamin content that wins the compliments at the dinner table. This attitude on the part of the diners is not as short sighted as it may at first appear. The sense organs concerned in the detection of taste and aroma (here collectively referred to as flavour) evolved to perform an essential function, to establish what was and what was not suitable as food. As a general rule palatability can be equated with nutritional value and most of the exceptions occur in relation to excess. For example, to the genuinely hungry, sweetness is a clear indicator of the presence of energy yielding sugars whereas to the mother of a well fed European child sweetness implies dental caries and obesity. Conversely, bitterness and astringency will prevent us from eating many poisonous plants, but are still sought after as elements in the flavour of beer and tea.

We usually regard taste as the property of liquids, or solids and gases in solution, that is detected in the mouth, not only by receptor cells in the taste buds of the tongue but also elsewhere in the oral cavity. Aroma is similarly regarded as the property of volatile substances detected by the receptor cells of the olfactory systems of the nose. Very few flavours, and fewer complete foodstuffs, allow a clear distinction to be made between an aroma and a taste and no attempt to discriminate absolutely between them will be made here.

The main interest in flavour for the food chemist is the identification of the particular substances responsible for particular flavour elements. One obvious motive for such work is that it enhances the possibilities for the simulation of natural flavours for use in processed food products. On occasions such investigations have made possible the identification of the features of chemical structure that invoke common response. An example is the identification of the 'AH,B' system involved in sweeteners which was discussed in Chapter 2.

The research chemist faces a number of special difficulties in investigations of flavour. The first problem is that there is no physical or chemical probe that can be used for the specific detection of the substances of interest. There is no flavour equivalent of the spectrophotometer which quantifies the light-absorbing properties of substances that we perceive as colour. Here the success of an analytical procedure may depend on the skill of a trained taste panel or the nose of an experimenter. Another complication is that a food flavour rarely depends on a single substance. It is commonly found that when the mixture of substances involved in a particular flavour has apparently been elucidated and then reconstituted in the 'correct' proportions one still does not have the flavour one started with. Usually this is because some substances, present in exceedingly small proportions, without strong flavour on their own and therefore overlooked by the analyst, have great influence on the total flavour of the food. These difficulties will not be underestimated if it is realised just how sensitive the nose can be. For example vanillin [6.1], the essential element of vanilla flavour can be detected by most individuals at a level of 0.1 ppm.

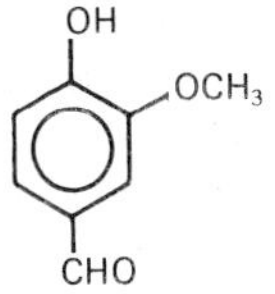

OH
OCH_3
CHO

6.1

Gas chromatography (GC) has proved to be the most valuable analytical technique available to the flavour chemist. Some of the most sophisticated modern instruments are capable of detecting as little as 10^{-14} g of a component in a mixture and 10^{-10} g is within the capabilities of most commercially available gas chromatographs. The peaks on a gas chromatogram can only be tentatively identified by comparison of elution characteristics with those of standard known substances. Firm identification requires that substances emerging from the chromatography column are separately trapped (usually by condensation at very low temperature) so that they can be subsequently identified by techniques such as infrared spectroscopy, nuclear magnetic resonance, and mass spectrometry. The vast amounts of information that sophisticated GC methods may make available do not automatically make the food chemist much wiser. For example the recent study of Scotch whisky in which 313 different volatile compounds were identified including 32 alcohols and 22 esters, will be slow to yield conclusions of real value to the manufacturers.

Although it is unwise to attempt an unequivocal distinction between tastes and odours or to suggest that a substance is involved in one but not the other it is usual to identify four major flavour sensations with the tongue and other parts of the mouth - saltiness, sweetness, sourness and bitterness. The mouth is also the major site of the responses to astringency, pungency and 'meatiness'. The substances involved in all these sensations have in common a number of characteristics that distinguish them from substances commonly associated with odours. Taste substances are usually polar, water soluble and non-volatile. Besides their necessary volatility odour substances are generally far less polar and elicit a much broader range of flavour sensations.

Saltiness is most readily detected on the sides and tip of the tongue and is elicited by many inorganic salts besides sodium chloride. However, the taste of salts other than sodium chloride has very little relevance to food. The contribution of sodium chloride to food flavour is often underestimated. We frequently add salt to both meat and vegetables to enhance their flavour but the sodium ions naturally present in many foods have a major role in their flavour. For example when sodium ions were eliminated from a mixture of amino acids, nucleotides, sugars, organic acids, and other compounds known to mimic the flavour of crab meat successfully (as well as corresponding closely to the composition of water soluble components of crab meat), the mixture was totally lacking in crab-like character. Besides this tendency to stimulate meaty flavours, salt tends to decrease the sweetness of sugars to give the richer, more rounded flavour that is required in many confectionery products.

Sweetness is detected by taste buds at the tip of the tongue when sugars and most other sweet-tasting substances are involved but there are some substances, notably the dihydrochalcones* (*eg* neohesperidin dihydrochalcone [6.2]) whose sweetness is detected at the back of the tongue. The contribution of sugars to the sweetness of foods was discussed at length in Chapter 2 and need not be repeated here.

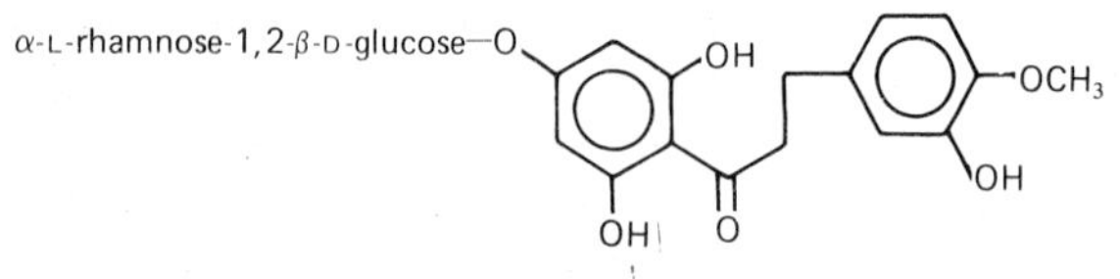

α-L-rhamnose-1,2-β-D-glucose—O ... OH ... OCH_3 ... OH ... OH ... O

6.2

Sourness is always assumed to be a property of solutions of low pH but it appears that H_3O^+ is much less important for taste than the undissociated forms of the organic acids that occur in acidic foodstuffs. In most fruit and fruit juices, citric acid [6.3] and malic acid [6.4] account for almost all the acidity. Tartaric acid [6.5] is characteristic of grapes; blackberries are especially rich in isocitric acid [6.6]; and rhubarb is of

*These dihydrochalcones are closely related to the *bitter* flavanoids such as naringin (see p. 133) which are found in some fruit juices.

course well known for its possession of oxalic acid [6.7] which occurs together with substantial amounts of malic and citric acids. Typical organic acid contents of citrus and grape juices are shown in Table 6.1.

```
   H2C—COOH        COOH        COOH      H2C—COOH
      |             |           |          |
HO—C—COOH       H—C—OH      H—C—OH     H—C—COOH     COOH
      |             |           |          |           |
   H2C—COOH        CH2      H—C—OH     H—C—COOH     COOH
                    |           |          |
                   COOH        COOH        OH

     6.3            6.4         6.5         6.6        6.7
```

Ethanoic (acetic) acid, at levels around 10–15 per cent, gives vinegar and products derived from it their sourness. In many pickled products such as pickled cabbage (*sauerkraut* in Germany) and cucumbers it is lactic acid, derived from the sugar in the vegetable by bacterial fermentation, that is responsible for the low pH and sourness. Lactic acid derived from lactose by fermentation also occurs in cheese at around 2 per cent where it contributes some of the sharpness.

Table 6.1. The organic acid content of fruit juices.

	Acid content (mM)		
	Malic	Citric	Tartaric
Orange	13	51	–
Grapefruit	42	100	–
Lemon	17	220	–
Grape	7	16	80

These values here are typical; wide variations occur between different varieties, degrees of ripeness, *etc*.

Bitterness is associated with several distinct classes of chemical substances. At the back of the tongue there are taste buds which are responsive to the bitterness of certain inorganic salts and also phenolic substances. Studies with the alkali metal halides have suggested that the structural criterion that distinguishes saltiness from bitterness is simply size. Where the sum of the ionic diameters is below that of KBr (0.658 nm) which tastes both salty and bitter, then the salty taste predominates, NaCl, at 0.556 nm, is an obvious example. KI, at 0.706 nm, tastes bitter. Magnesium salts such as $MgCl_2$ (0.850 nm) also have a bitter taste. Phenolic substances in the form of flavanoids are important sources of bitterness in fruit juices – particularly citrus juices. The best known is naringin [6.8], a glycoside of the flavanone naringenin with the disaccharide neohesperidose, which occurs in grapefruit and Seville oranges. Its bitterness is such that it can be detected at a dilution of 1 in 50 000. Another bitter element that can occur in citrus

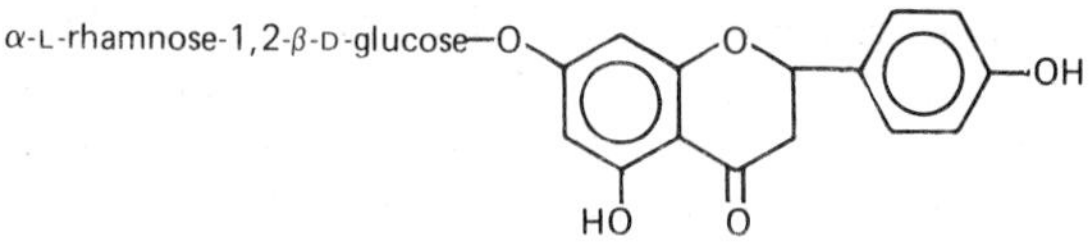

α-L-rhamnose-1,2-β-D-glucose—O

6.8

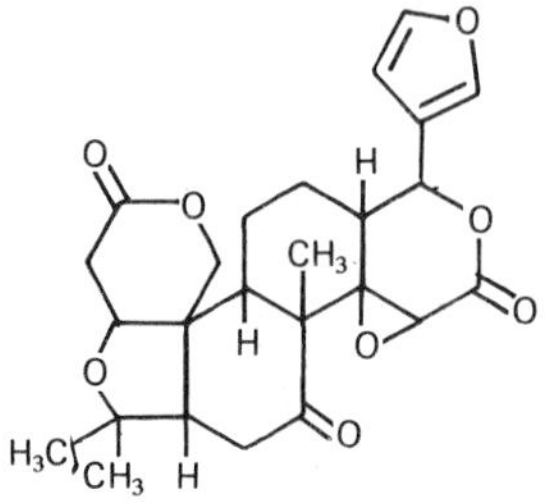

6.9

juices is limonin [6.9], which is sometimes formed from tasteless precursors during commercial juice extraction.

Bitterness is a sought after characteristic in many types of beer, particularly those brewed in Britain. Bitterness is achieved by adding hops to the *wort*, *ie* the sugary extract from the malt, before it is boiled and then cooled in the stage preceding the actual fermentation. Hops are the dried flowers of the hop plant *Humulus lupulus* and they are rich in volatile compounds that give beer its characteristic aroma together with resins that include the bitter substances. The most important bittering agents are the so-called α-acids. As shown in *Fig. 6.1* these isomerise during the 'boil' to give forms which are much more soluble in water and much more bitter. Different varieties of hops have different proportions of the three α-acids and the structurally related, but less important, β-acids.

The ability to perceive bitterness almost certainly evolved in order to protect man, and presumably other animals, from the dangers posed by the alkaloids present in many plants. Alkaloids are defined as basic

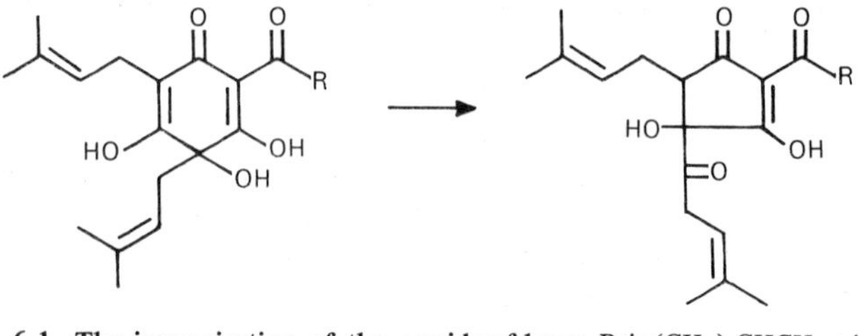

FIG. 6.1. The isomerisation of the α-acids of hops. R is $(CH_3)_2CHCH_2$— in humulone, $(CH_3)_2CH$— in cohumulone and $C_2H_5(CH_3)CH$— in adhumulone.

organic compounds having nitrogen in a heterocyclic ring. Although we have little idea as to the function of the alkaloids in plants, many have highly undesirable pharmacological effects on animals besides their extremely bitter taste; nicotine, atropine and emetine are all alkaloids. In view of its medicinal use quinine [6.10] is one of the best known and it is much used as a bittering agent in soft drinks such as 'bitter lemon' and 'tonic water'.

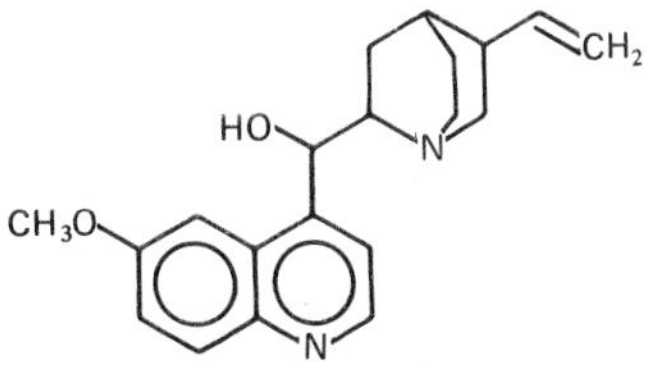

6.10

Until recently the bitter taste of many amino acids and oligopeptides was of only academic interest. However, the need for more efficient use of the proteins in whey, waste blood *etc* has focused attention on the properties of the peptides that are obtained by partial hydrolysis, by enzymes or acid, of these proteins. These protein hydrolysates are being developed as highly nutritious food additives with many useful properties relating to food texture. Examination of the tastes of the amino acids that occur in protein hydrolysates shows that bitterness is exclusively the property of the hydrophobic L-amino acids – valine, leucine, isoleucine, phenylalanine, tyrosine and tryptophan. Whether a peptide is bitter or not depends on the average hydrophobicity of its amino acid residues. A great deal of work on the conformation of peptides, in relation to their taste, has demonstrated that the structural requirements for bitterness mirror those required for sweetness in sugars and other compounds. There is a similar requirement for a correctly spaced pair of hydrophilic groups (AH and B in sugars), one basic and one acidic, together with a third, hydrophobic group in the correct spatial relationship to the hydrophilic groups. The ability to predict the likely taste of a peptide can be linked to our knowledge of amino acid sequences to predict the appearance of bitter, and therefore unwelcome, peptides in hydrolysates of particular proteins. For example caseins and soya proteins are rich in hydrophobic amino acids and therefore tend to yield bitter peptides. One possible solution is to ensure that the hydrolysis is limited to producing fragments with molecular weights above about 6000 which are too large to interact with the taste receptors.

Astringency is a sensation that is clearly related to bitterness but is registered within the oral cavity generally as well as on the tongue. Astringency is usually regarded as a desirable characteristic of fruit and cider but it is in red wine and tea that it is most important. In both these beverages it is associated with the high content of the polyphenolic substances, which are also involved in colour, and were described in Chapter 5. In wine and tea the polyphenols responsible in part for flavour are collectively described as *tannins*. In black tea most of the polyphenolic material is considered to contribute to the overall astringency* but it is the gallyl groups of the theaflavin gallates and other polyphenols that are most important. There is evidence for an interaction between these groups and the caffeine (which alone is merely bitter) in producing astringency. In wine the situation is more complex as there are many more elements in the flavour, but all the evidence suggests that it is flavanoids similar to those found in tea that are responsible for the astringency. Besides their obvious anthocyanin content, which does contribute a little to the taste, red wines generally have up to 800 mg dm^{-3} of catechins (as in *Fig. 5.10*) compared with no more than 50 mg dm^{-3} in white wine.

Pungency is another sensation experienced by the entire oral cavity. It is the essential characteristic of a number of important spices but it is also found in many members of the family *Cruciferae*. The genus *Capsicum* includes the red and green chillies, the large *C. annum* being much less pungent than the ferocious (as far as I am concerned this is the only word that does them justice) little fruit of *C. frutescens*. The active components in *Capsicum* species are known as capsaicinoids, capsaicin [6.11] and dihydrocapsaicin (which has a fully saturated side chain) being the most abundant and the most pungent.

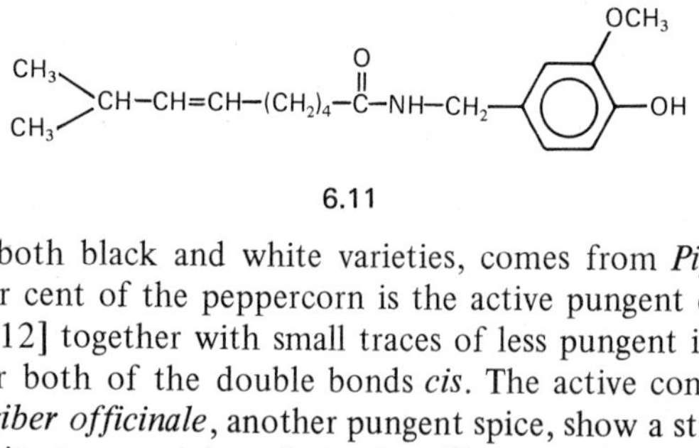

$$(CH_3)_2CH{-}CH{=}CH{-}(CH_2)_4{-}\overset{O}{\overset{\|}{C}}{-}NH{-}CH_2{-}C_6H_3(OCH_3)(OH)$$

6.11

Pepper, both black and white varieties, comes from *Piper nigrum*. About 5 per cent of the peppercorn is the active pungent component, piperine [6.12] together with small traces of less pungent isomers that have one or both of the double bonds *cis*. The active components of ginger, *Zingiber officinale*, another pungent spice, show a striking structural similarity to capsaicin and piperine. These are the gingerols [6.13] and shogaols [6.14]. Gingerols and shogaols with *n* equal to 4 predominate but higher homologues do occur in small amounts. Different forms of ginger, *eg* green root ginger and dry powdered ginger, have different proportions of gingerols and shogaols as a result of the readiness with which gingerols dehydrate to the corresponding shogaols. The vanillyl side chain is also found in eugenol [6.15], the essential pungent component of cloves.

*The astringency of tea is often referred to as 'briskness' by tea experts.

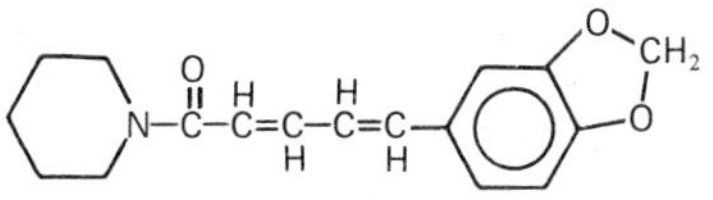

6.12

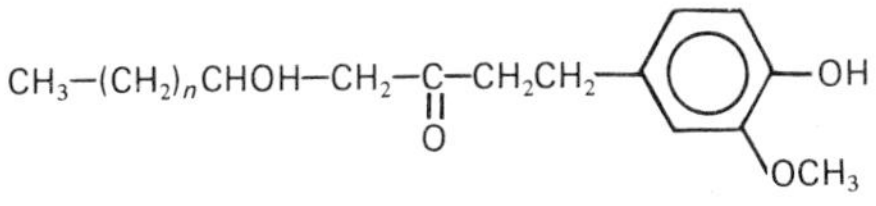

6.13

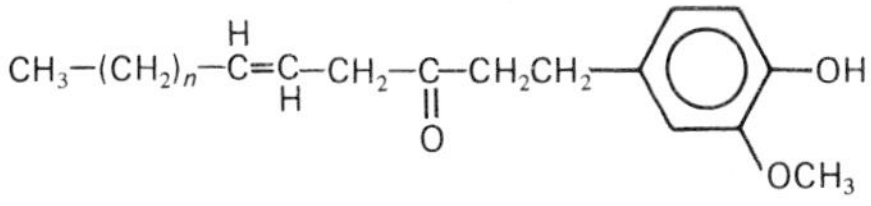

6.14

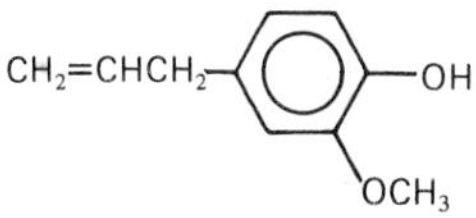

6.15

The final group of pungent food components to be considered, those of the family *Cruciferae*, have no structural features in common with these spice components. This family contains a number of plants renowned for their pungency when raw, such as horseradish (*Amoracia lapathiofolia*), black and white mustards (*Brassica nigrum* and *B. alba*) and radish (*Raphanus sativus*). A lesser degree of pungency is also found in the raw tissues of other leafy vegetables of the *Brassica* genus such as cabbage, brussels sprouts and kale. The cells of these plants contain members of a group of compounds known as *glucosinolates* (see *Fig. 6.2*). When the tissues of these plants are physically disrupted, by biting or chewing in the mouth or by cutting, grating *etc* in the kitchen the enzyme *myrosinase* is liberated from specific cells and catalyses their breakdown. The immediate products of the enzyme action, thiohydroxamate-*O*-sulphates, spontaneously lose the sulphate and isomerise to give the volatile and pungent isothiocyanates and a number of other products. Whenever the vegetable is cooked, and the enzyme destroyed before tissue disruption brings the enzyme and substrate into contact, the isothiocyanates are not formed. Instead a wide range of other sulphur compounds arise which give over-boiled cabbage and other similar vegetables their characteristic unattractive odour. The difference in flavour between frozen and fresh brussels sprouts after cooking has been shown to reflect the different rates of myrosinase destruction during commercial blanching and domestic cooking. Nitrile

glucosinate

$\xrightarrow{H_2O}$ Glucose

$R{-}C(SH){=}N{-}OSO_3^-$ thiohydroxamate-*O*-sulphate

$R{-}N{=}C{=}S$ isothiocyanates

$R{-}C{\equiv}N$ nitrile

$R{-}S{-}C{\equiv}N$ thiocyanate

The identities of R in the predominant glucosinolates of a number of vegetables:

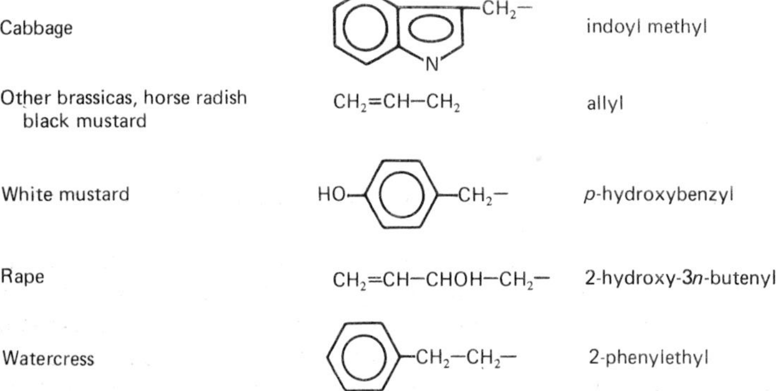

FIG. 6.2. The breakdown of glucosinolates to form isothiocyanates and other products. The side chains (R–) of the predominant glucosinolates of various plants are shown.

formation is a significant problem in the utilisation of rape seed meal, *ie* the residue from rape seed oil extraction. Another problem with rape seed meal is the tendency for the isothiocyanate to cyclise spontaneously to a compound [6.16], which is goitrinogenic, *ie* interferes with iodine uptake by the thyroid gland. In spite of this the suggestion that milk from cattle fed largely on rape seed and other brassicas might be harmful should not be taken too seriously.

$$CH_2{=}CH{-}CHOH{-}CH_2{-}N{=}C{=}S \longrightarrow CH_2{=}CH{-}\underset{\displaystyle H_2C{-}NH}{CH{-}O{-}C{=}S}$$

6.16

Although the adjective pungent may not be entirely appropriate this is nevertheless an appropriate point at which to mention the flavour compounds of onion, *Allium cepa* and garlic, *A. sativum*. All members of this genus contain *S*-alkyl cysteine sulphoxides (see *Fig. 6.3*), com-

$$R{-}\underset{\substack{\|\\O}}{S}{-}CH_2{-}\underset{\substack{|\\NH_2}}{CH}{-}COOH \xrightarrow[\substack{\downarrow\\CH_3\\|\\C=O\\|\\COOH}]{H_2O \quad NH_3 \uparrow} R{-}\underset{\substack{\|\\O}}{S}H \quad \text{unstable alkyl sulphenic acid}$$

S-allyl cysteine sulphoxide

(*i*) Onion: The 1-propenyl sulphenic acid spontaneously isomerises to the lachrymatory thiopropionaldehyde-*S*-oxide

$$CH_3{-}CH{=}CH{-}\underset{\substack{\|\\O}}{S}H \longrightarrow CH_3CH_2CH{=}S{=}O$$

(*ii*) Garlic: The allyl sulphenic acid spontaneously dimerises to give allicin (diallylthiosulphinate) which has antibacterial properties

$$2CH_2{=}CH{-}CH_2{-}\underset{\substack{\|\\O}}{S}H \longrightarrow CH_2{=}CH{-}CH_2{-}\underset{\substack{\|\\O}}{S}{-}S{-}CH_2{-}CH{=}CH_2$$

FIG. 6.3. The breakdown of *S*-alkyl cysteine sulphoxides in onion and garlic.

monly the methyl propyl derivatives but the allyl derivative is particularly associated with garlic.

The 1-propenyl derivative is similarly associated with onions. Just as in isothiocyanate formation the compounds that occur in the raw, undamaged plant tissues are quite odourless. It is only when the garlic is crushed or the onion is sliced that the enzyme, *alliinase*, gains access to its substrate and the flavour becomes obvious. During cooking the thiopropionaldehyde-*S*-oxide undergoes a variety of reactions involving sulphur loss, dimerizations, and condensations with other cell components. These can lead to compounds which are important in the odour of the cooked onion, such as 2-methyl-but-2-enal and 2-methylpent-2-enal. The 'sulphury' odours of cooked onions and garlic arise from the products of complex disulphide exchange reactions between cysteine and dialkylthiosulphinates:

$$R{-}\underset{\substack{\|\\O}}{S}{-}S{-}R \;+\; HS{-}CH_2\underset{\substack{|\\NH_2}}{C}HCOOH \longrightarrow R\underset{\substack{\|\\O}}{S}H \;+\; R{-}S{-}S{-}CH_2\underset{\substack{|\\NH_2}}{C}H{-}COOH$$

$$R{-}S{-}S{-}CH_2\underset{\substack{|\\NH_2}}{C}H{-}COOH \xrightarrow{\times 2} R{-}S{-}S{-}R \;+\; \text{cystine}$$

Of all the taste classes the most recent to attract attention is meatiness. Although the flavour of meat is obviously an amalgam of our responses to both volatile and non-volatile compounds, meaty taste has come to be associated with two particular substances that actually occur in many other foods besides meat, inosine monophosphate, IMP [6.17] and monosodium-L-glutamate (MSG).

Alone, neither of these substances has a particularly strong taste. They need to be at a concentration above about 300 mg dm^{-3} before they make much impact. However, when present together they show a remarkable synergism. A mixture of equal proportions of IMP and MSG tastes some 20 times stronger than the same total amount of one alone. Although the mechanism of this synergism is not understood, its importance to food manufacturers is obvious. The addition of only a small amount of one of these substances (usually MSG) to a food product that naturally contains modest amounts of the other will have a dramatic effect on the flavour. Thus MSG is used as a *flavour enhancer* rather than flavouring in dried soups and similar products, at levels to give about 1 g dm^{-3} in the final dish. Although they are much more expensive, preparations of IMP, often blended with guanosine monophosphate [6.18], are now being used by some food manufacturers.

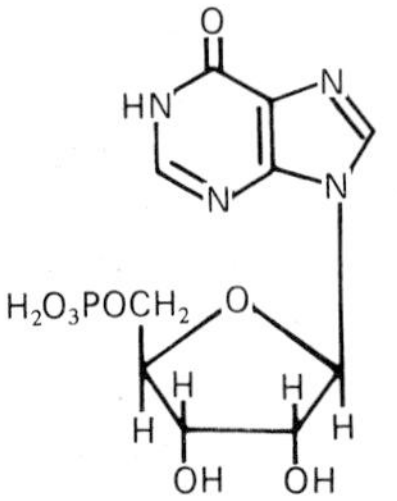

6.17

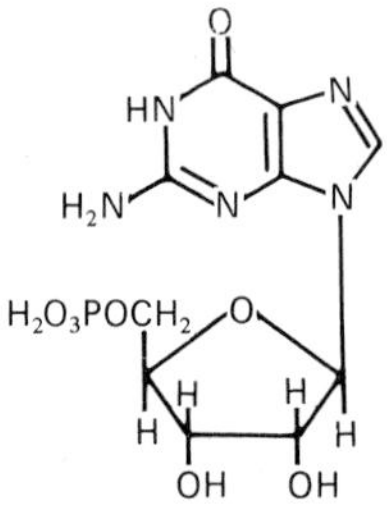

6.18

Besides GMP, which does occur naturally in animal tissues, there are numerous highly unnatural derivatives and analogues of IMP that also have a meaty taste, often stronger than IMP, but these are only of academic interest. Some amino acids are also said to have a weak meaty taste but their effect on food flavour appears to be only marginal. Glutamic acid, as MSG, will naturally be present at low levels in muscle tissue but in meat it will be derived mostly from the protein breakdown that occurs during the ageing process. IMP is a breakdown product of the adenosine monophosphate, AMP, that accumulates as ATP is utilized in the muscle *post-mortem*:

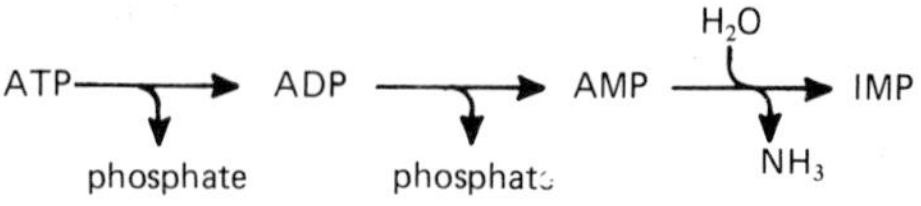

$$\text{ATP} \xrightarrow[\text{phosphate}]{} \text{ADP} \xrightarrow[\text{phosphate}]{} \text{AMP} \xrightarrow[\text{NH}_3]{\text{H}_2\text{O}} \text{IMP}$$

While IMP and MSG provide the basic 'meatiness' of meat and meat products the subtle differences in flavour between different meats depend on variations in their proportions and also involve many other flavour compounds that occur in trace amounts. For example the IMP contents of beef and pork are roughly similar (100–150 mg per 100 g) but beef has significantly more free amino acids than pork, including twice as much MSG. Lamb is especially rich in MSG but has markedly low levels of dipeptides such as carnosine [6.19] which also contribute to the meat flavour. An increase in the level of free amino acids with maturity has been suggested as the difference between the flavour of veal and beef.

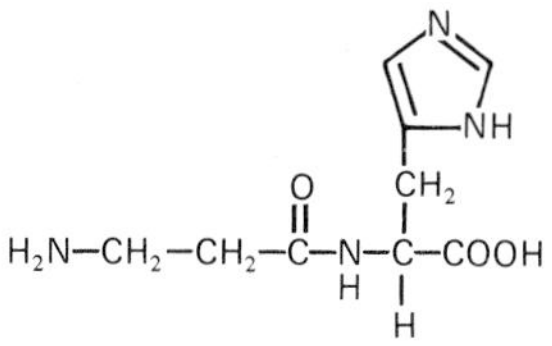

$$H_2N-CH_2-CH_2-C(=O)-NH-CH(CH_2\text{-imidazolyl})-COOH$$

6.19

Although some dishes do depend on raw meat (*eg* steak tartare) cooking is obviously the origin of the innumerable volatile compounds that are so important in our appreciation of meat. The vast numbers of novel compounds that have been isolated defy systematic discussion but certain classes of compounds have been identified in many cooked meat systems. Thermal decomposition of ribonucleotides such as IMP has been shown to lead to compounds such as methyl furanolone [6.20] which not only has a pronounced meaty odour itself but can also be a precursor of sulphur-containing compounds (the sulphur is derived from the breakdown of cysteine and methionine) such as methyl thiophenone [6.21] which also have a strong meaty odour.

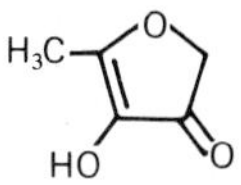

6.20 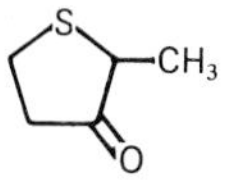 6.21

Fatty acid breakdown is also an important source of the characteristic odours in meat products. The arachidonic acid (see Chapter 3) that occurs in the polar lipids of muscle tissue has been shown to break down on heating, presumably by oxidation reactions similar to those described in Chapter 3, to give four aldehydes which together have a distinctive cooked chicken odour:

$CH_3(CH_2)_4CH{=}CHCH_2CHO$	3-*cis*-nonenal
$CH_3(CH_2)_4CH{=}CH(CH_2)_2CHO$	4-*cis*-decenal
$CH_3(CH_2)_4CH{=}CHCH_2CH{=}CHCHO$	2-*trans*,5-*cis*-undecadienal
$CH_3(CH_2)_4CH{=}CHCH_2CH{=}CHCH{=}CHCHO$	2-*trans*,4-*cis*,7-*cis*-tridecatrienal

Many odorous flavour compounds have now been mentioned in this or earlier chapters but they require special attention when the flavours of fruit are considered. The *tastes* of most fruit are a blend of sweetness due to sugars (mostly mixtures of glucose, fructose and sucrose) and the sourness of organic acids such as citric and malic. However, to distinguish between fruit flavours we rely heavily on the distinctive odours of the fruit's volatile constituents. For example, with a cold it is extremely difficult to distinguish between raspberry and strawberry flavours.

A typical fruit may well have as many as 200 different volatile components but even in total these may comprise only a few parts per million of the total fruit. Long lists of compounds have been assembled for many fruit which show that acids, alcohols, the esters resulting from their combination, aldehydes and ketones predominate *numerically*. Considering apples as a fairly typical example we find:

at least 20 aliphatic acids ranging from formic to *n*-decanoic;

at least 27 aliphatic alcohols covering a corresponding range of chemical structures;

over 70 esters involving the predominant alcohols and acids;

26 aldehydes and ketones also with structures corresponding to the acids and alcohols;

smaller numbers of miscellaneous ethers, acetals, terpenoids and other hydrocarbons,

a total of 131! In a consideration of aroma the first point to remember is that by no means all of these substances make a significant contribution; volatility does not equal aroma. Another point is that a very high proportion of these compounds are common to many different fruits. For example, of the 17 esters identified in banana volatiles only five are not found in apples. The particularly common occurrence of aldehydes, alcohols and acids with linear or branched (*eg* isopropionyl) chains of up to seven carbon atoms, often with an occasional double bond, suggests common origins. In fact there are two sources which are both represented in many fruits. During ripening, cell breakdown is accompanied by the oxidation of the unsaturated fatty acids of membrane lipids catalysed by the enzyme *lipoxygenase*. This converts fatty acids with a *cis,cis* methylene interrupted diene system (as in linoleic acid) to hydroperoxides. These break down spontaneously or under the influence of other enzymes to give aldehydes. The pattern of reaction products is very similar to that found in the non-enzymic autoxidation reactions of fatty acids. If tender vegetables such as peas and beans are not rapidly blanched to destroy enzyme activity soon after harvest the accumulation of these aldehydes leads to undesirable off-flavours.

Amino acids are the other sources of aldehydes. Enzymic transamination and decarboxylation of free amino acids often accompanies ripening. For example leucine gives rise to 3-methylbutanal:

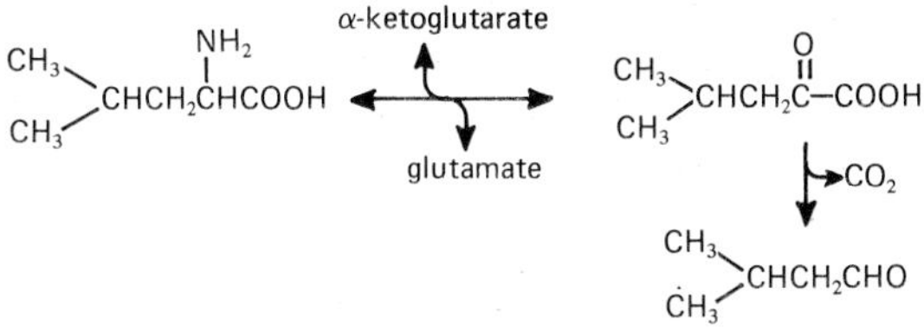

$$(CH_3)_2CHCH_2CH(NH_2)COOH \xrightleftharpoons[\text{glutamate}]{\alpha\text{-ketoglutarate}} (CH_3)_2CHCH_2C(=O){-}COOH \xrightarrow{-CO_2} (CH_3)_2CHCH_2CHO$$

The aldehydes are not only important in fruit and vegetables. During alcoholic fermentations, yeast enzymes catalyse similar reactions with the amino acids from the malt or other raw materials. These aldehydes, and the alcohols derived from them, are important as flavour compounds in many alcoholic drinks, particularly distilled spirits (where they are described as *fusel oils*), and where they have a special responsibility for the headaches associated with hangovers. Other enzyme systems in plant tissues are responsible for the reduction of the aldehydes to alcohols or their oxidation to carboxylic acids, and the combination of these to form esters.

In many cases the distinctive character of a particular fruit flavour is dependent on one or two fairly unique 'character impact' substances. For example, isopentyl acetate is the crucial element in banana flavour although eugenol [6.15] and some of its derivatives contribute to the mellow, full bodied aroma of ripe bananas. Benzaldehyde is the 'character impact' substance in cherries and almonds. It is difficult to establish to what extent HCN, which is also present at low levels (a few ppm at most) in cherries contribute to their aroma. Hexanal and 2-hexenal provide what are described as 'green' or 'unripe' odours in a number of fruits and vegetables. In at least one variety of apple they have been shown to balance the 'ripe' note in the aroma given specifically by ethyl 2-methyl butyrate. The olfactory thresholds of these three substances are, respectively 0.005, 0.017 and 0.0001 ppm! The distinctive character of raspberry aroma is due mostly to 1-(*p*-hydroxyphenyl)-3-butanone [6.22] but the fresh grassy aroma which is characteristic of the fresh fruit, and notably lacking in many raspberry flavoured products, is principally contributed by *cis*-3-hexenol supplemented by α- and β-ionone (see Chapter 5).

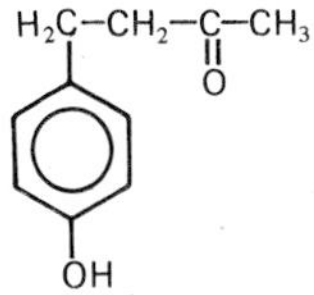

$$HO{-}C_6H_4{-}CH_2{-}CH_2{-}C(=O){-}CH_3$$

6.22

The character impact substances in many herbs and spices and especially citrus fruit belong to the class of organic molecules known as 'terpenoids'. Terpenoids are defined as naturally occurring hydrocarbons (the terpenes) and their oxygenated derivatives, which are based structurally and biosynthetically on the condensation of isoprenoid units. These have five carbon atoms arranged:

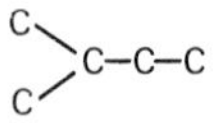

C
>C–C–C
C

The terpenoids are classified according to the number of isoprenoid units they contain - monoterpenoids having two, diterpenoids having four and so on. We have already encountered some terpenoids in Chapter 5; the phytol side chain of chlorophyll is a diterpenoid, the carotenoids are tetraterpenoids.

Terpenoids are isolated from plant materials by means of steam distillation. The oily fraction of the distillate is known as the essential oil. The essential oils of citrus fruits (*ie* of the entire fruit including the peel) consist mainly of terpenes, of which the monoterpene (+)-limonene [6.23] is usually at least 80 per cent. However, (+)-limonene is not nearly so important as a flavour component as the oxygenated terpenoids which occur in the oil in much smaller amounts. For example fresh grapefruit juice contains about 16 ppm of (+)-limonene, carried over from the peel during juice preparation, but the characteristic aromatic flavour of grapefruit is due to the presence of much smaller amounts of nootkatone [6.24]. The characteristic bitterness of grapefuit juice was mentioned earlier in this chapter. The character impact substance in lemons is citral, which is more correctly named as a mixture of the isomers geranial [6.25] and neral [6.26]. Oranges do not appear to possess a clearly identified character impact substance.

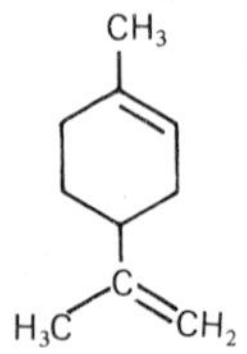

CH_3 ... H_3C C CH_2 — 6.23

O ... CH_3 CH_3 CH_2 CH_3 — 6.24

CH_3 CHO H_3C CH_3 — 6.25

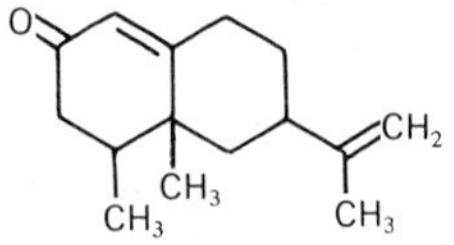

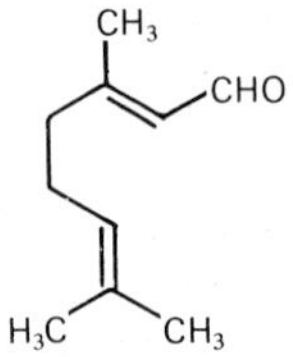

CH_3 CHO H_3C CH_3 — 6.26

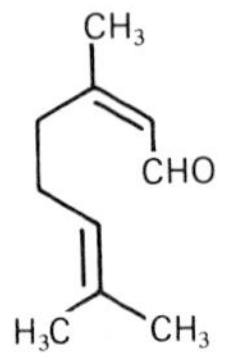

A principal objective for the flavour chemist is the provision of synthetic flavourings for use by both the housewife and the food manufacturer. An important trend in recent years has been the replacement of simple esters, with their sweet and unsubtle aromas (*eg* amyl acetate in pear drops) with lighter, fresher and very much more expensive flavourings based more closely on our knowledge of the natural flavour

Table 6.2. Synthetic fruit flavourings.

	Parts
Raspberry (from US Patent No 3886289)	
Vanillin	20
Ethylvanillin	8
α-Ionone*	1
Maltol	30
1-(*p*-hydroxyphenyl)-3-butanone	100
Dimethyl sulphide	1
2,5-Dimethyl-*N*-(2-pyrazinyl)pyrrole*	1
Strawberry (from US Patent No 3686004)	
Geraniol	1
Ethyl methyl phenyl glycidate	3
2-Methyl-2-pentenoic acid	5
Vanillin	6
Ethyl pelargonate	13
Isoamyl acetate	14
Ethyl butyrate	52
1-(prop-1-enyl)-3,4,5-trimethoxybenzene*	1
Pineapple	
Ethyl butyrate	60
Isoamyl butyrate	20
Allyl caprate	5
Glycerol	5
Lemon oil	1
Ethyl acetate	1

Table 6.3. A synthetic chocolate flavouring (US Patent 3619210).

	Parts
Dimethyltrisulphide	1
2,6-Dimethylpyrazine	3324
Ethylvanillin	143
Isovaleraldehyde	100

components. Very little information on the composition of their products is ever released by flavour manufacturers but the patent literature does give some insights. This chapter is concluded with some proposed formulations (Tables 6.2 and 6.3) for use in confectionery, soft drinks and other products. For simplicity the proportions of carriers, solvents *etc* have been omitted. The patented raspberry and strawberry formulations both contain components (marked with an asterisk) which are considered to add fresh or woody notes to the flavour to give greater authenticity. In recent years a demand has built up in the food industry for flavourings outside the range of the traditional fruit flavours. For example the very high price of cocoa has stimulated demand for good quality chocolate flavourings. The natural flavour of chocolate has been extensively but inconclusively investigated. Two groups of compounds have been implicated, sulphides (*eg* dimethyl sulphide and other sulphur compounds) and pyrazines. A total of 57 different pyrazines have been identified in cocoa volatiles! A patented chocolate flavouring formulation, containing a representative pyrazine and sulphide is given in Table 6.3. In view of the understandable desire of most people to minimise the number of 'chemicals' in our food perhaps we should consider the replacements of natural ingredients with synthetic ones!

Further reading

Phenolic sulphur and nitrogen compounds in food flavours (G. Charalombous and I. Katz eds). Washington: American Chemical Society, 1976.

The quality of foods and beverages (G. Charalombous and G. Inglett eds), Vol. I. New York: Academic Press, 1981.

Food taste chemistry (J. C. Boudreau ed). Washington: American Chemical Society, 1979.

R. Teranishi *et al*, *Flavour research: principles and techniques*. New York: Marcel Dekker, 1971.

Food flavours; part A: introduction (I. D. Morton and A. J. Macleod eds). Amsterdam: Elsevier, 1982. (This excellent, if very expensive, work includes articles on flavours derived from carbohydrates and lipids which are dealt with in Chapters 2 and 3 of this book.)

N. D. Pintauro, *Food flavouring processes*. Park Ridge, New Jersey: Noyes Data Corporation, 1976.

The biochemistry of fruits and their products (A. C. Hulme ed). London: Academic Press, 1970.

7. Vitamins

Accounts of the discovery of the existence of vitamins, their occurrence and their importance in the diet are to be found in almost every textbook of nutrition, biochemistry or physiology and need not be repeated here. Our primary concern is their chemistry. The first task for chemists was the isolation of individual vitamins in a pure form and the determination of their chemical structures. Once their structures were known chemists were called upon to provide sensitive and accurate methods for the determination of vitamin levels in food materials. Initially only biological assays had been available, based on the relief of deficiency diseases in laboratory animals, but their high cost and great slowness made them unsuitable for routine measurements.* As we shall see this call is yet to be fully answered. A second objective for vitamin chemists has been the provision of commercially viable chemical syntheses of vitamins for use as dietary supplements. Only in the case of ascorbic acid (vitamin C) has this proved successful so far.

The vitamins are an untidy collection of complex organic substances that occur in the biological materials we consume as food (and those we do not). In terms of chemical structure they have nothing in common and their biological functions similarly offer no help in their definition or classification. What does draw them together is that they:

(*i*) tend to occur in only tiny† amounts in biological materials
(*ii*) are essential components of the biochemical or physiological systems of animal life (and frequently plant and microbial life)
(*iii*) as animals evolved they lost the ability to synthesise these substances for themselves in adequate amounts.

---

*In the earlier years of this century the moral question of the use of laboratory animals was sadly less prominent than it is today.

†As distinct from the essential fatty acids and essential amino acids considered in earlier chapters.

Animals are distinguished from most other forms of life by their dependence on other organisms as food. Directly or indirectly plants are the fundamental source of basic nutrients and it is hardly surprising that animals have come to rely on plants for the supply of other substances as well. The risks of vitamin deficiencies occurring in an otherwise adequate diet would have been slight, when this dependence evolved, and a small price to pay for the loss of the burden of maintaining the synthetic machinery for such a diverse range of complex substances.

The naming of vitamins is somewhat problematical. Before there was any information as to the chemical structure of vitamins, systematic chemical names were clearly impossible - even today they are usually too cumbersome. Two alternatives evolved. One was to refer to the disease caused by a deficiency of the vitamin, *eg* anti-pellegra factor. The other, more systematic approach was to assign a letter of the alphabet as each vitamin was discovered - vitamin A, vitamin B *etc*. This system ran into trouble as soon as it was found that the original vitamin B was in fact a collection of numerous quite distinct vitamins. The result was the addition of numbers so that vitamins B_1, B_2 *etc* appeared. This system was finally reduced to chaos and abandoned when many differently lettered and numbered vitamins turned out to be identical. Thus there is now no trace of vitamins F to J, B_3 to B_5! Although fairly simple chemical names have now largely supplanted these earlier designations some still linger on. Most people now recognise that ascorbic acid is vitamin C but there is no sign that vitamins A and D are becoming widely known as retinol and cholecalciferol. The only element of vitamin classification that has persisted is applied to the accounts of individual vitamins that follow, *ie* a division of the list into the water soluble vitamins and the fat soluble ones. It will be seen that this split is to some extent related to both their occurrence and function.

Thiamine (vitamin B_1, aneurine)

Thiamine occurs in foodstuffs either in its free form [7.1] or as its pyrophosphate [7.1a] ester complexed with protein.

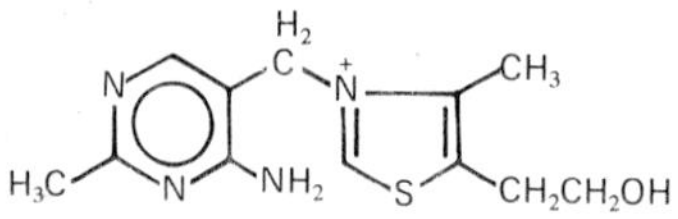

7.1

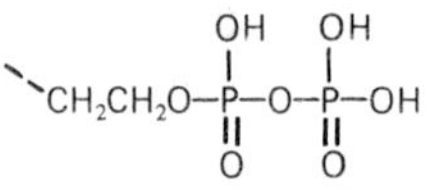

7.1a

No distinction is made between these forms by the usual analytical techniques or in tables of food composition. Although it is extremely widespread in small amounts only a few foodstuffs can be regarded as good sources. As a general rule it is present in greatest amounts (0.1–1.0 mg per 100 g) in foodstuffs that are rich in carbohydrate and/or associated with a high level of carbohydrate metabolism in the original living material. Examples are legume seeds and the embryo component of cereal grains (the germ) which are involved in high rates of carbohydrate metabolism during the germination of the seed. Although meat, both muscle tissue and liver, contains very little carbohydrate in the living animal, carbohydrate is the chief source of energy for these tissues. The reason why pork should contain about 10 times the amount of thiamine (about 1 mg per 100 g) that beef, lamb, poultry and fish contain has yet to be established.

The association of thiamine with carbohydrate is related to its role in metabolism. Thiamine pyrophosphate (TPP), is the prosthetic group* of a number of important enzymes catalysing the oxidative decarboxylation of α-keto acids including pyruvic [7.2] and α-ketoglutaric [7.3] which occur in the Krebs cycle.

$$CH_3\text{—}C(=O)\text{—}COOH$$

7.2

$$COOH\text{—}CH_2\text{—}CH_2\text{—}C(=O)\text{—}COOH$$

7.3

The role of TPP in oxidative decarboxylation is shown in *Fig. 7.1*. Although details, such as these, of enzyme reaction mechanisms are not normally the province of food chemists this is a valuable illustration of *why* vitamin structures are so often complex. It is also worth noting that the essential component of coenzyme A, pantothenic acid, is also a vitamin. There is some evidence that thiamine, probably as the pyrophosphate, has a quite separate role in nerve function, but details are not established.

Thiamine is one of the few water soluble vitamins that can be fairly easily determined by chemical methods and, even so, the method requires the use of sophisticated techniques. After extraction from the food material with hot dilute acid and treatment with the enzyme phosphatase to convert any TPP to thiamine the extract is 'cleaned-up' by column chromatography. This extract is then treated with an oxidising agent, usually hexacyanoferrate(II) (ferricyanide), to convert the thiamine to thiochrome [7.4] whose concentration is measured fluorimetrically.

---

*A prosthetic group is a non-protein component of an enzyme which is involved in the reaction catalysed and remains bound to the enzyme throughout the reaction sequence.

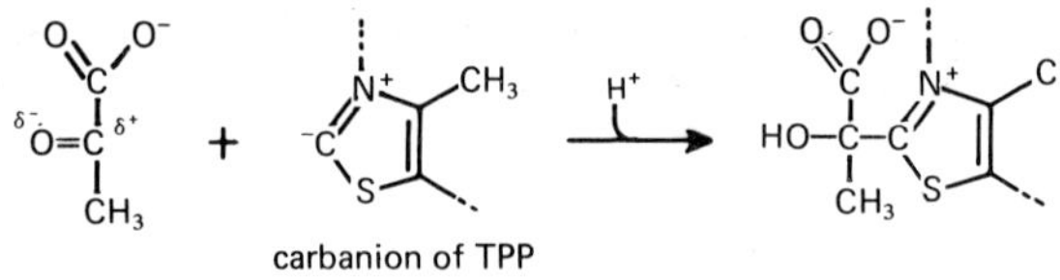

The carbon atom between the nitrogen and sulphur of the thiazole ring is highly acidic and ionizes to form a carbanion. This readily adds to the carbonyl group of the α-keto acid (pyruvate is the illustrated example).

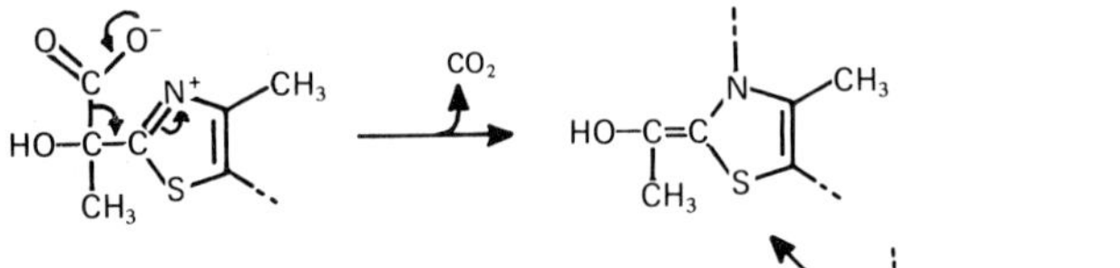

The positive charge on the nitrogen then facilitates the loss of CO_2. The hydroxyethyl group is then transferred, with concomitant oxidation, to the enzyme's second prosthetic group, lipoamide. From here it is finally transferred, as an acetyl group, to the sulphydryl group of coenzyme A.

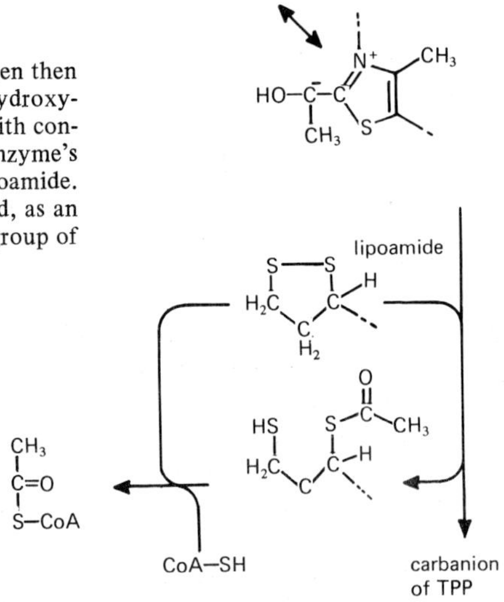

FIG. 7.1. The mechanism of oxidative decarboxylations involving thiamine.

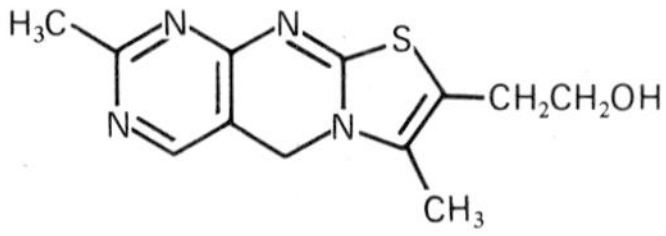

7.4

Thiamine is one of the most labile of vitamins. Only under acid conditions, *ie* below pH 5, will it withstand heating. The reason for this is the readiness with which nucleophilic displacement reactions occur at the atom joining the two ring systems. Both OH^- and, more seriously, HSO_3^-, split the molecule in two:

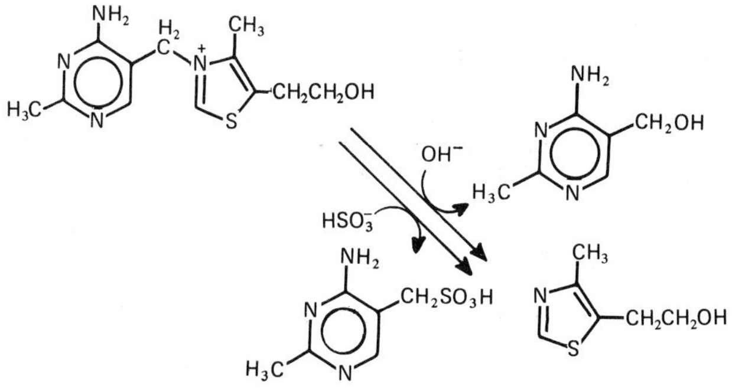

Thus sulphite added to fruits and vegetables to prevent browning will cause total destruction of the thiamine. Fortunately these foods are not important sources of dietary thiamine.

Heat treatments such as canning, especially when the pH is above 6, cause losses of up to 20 per cent of the thiamine but only in the baking of bread, where up to 30 per cent may be lost, are these losses really important.

The greatest losses of thiamine, in domestic cooking as well as commercial food processing, occur simply as a result of its water solubility rather than any chemical subtleties.

It is impossible to generalise on the extent of such losses, depending as they do on the amount of chopping up, soaking and cooking times *etc*, but they can be minimized by making as much use as possible of meat drippings and vegetable cooking water – perhaps gravy is nutritious after all!

Thiamine also provides us with a curious example of a nutritious advantage to be obtained from cooking. Raw, fermented fish, a popular food in many parts of the Far East such as Thailand, contains an enzyme, thiaminase, derived from the bacteria which carry out the fermentation. The destructive action of this enzyme can be enough to cause thiamine deficiency in diets that apparently contain adequate levels. Particularly in Thailand this problem is aggravated by the habit of continuously chewing tea leaves. The polyphenolic compounds extracted from the leaves combine with thiamine and thereby destroy its vitamin activity.

Riboflavin (vitamin B_2)

The structure of this vitamin is usually presented as that of riboflavin itself [7.5], the isoalloxazine nucleus with just a ribitol side chain

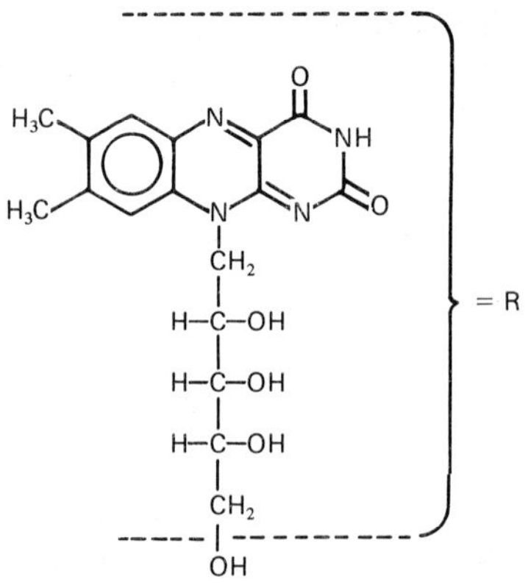

H_3C H_3C N NH O O N N CH_2 H–C–OH H–C–OH H–C–OH CH_2 OH = R

7.5

$R–OPO_3H_2$

7.6

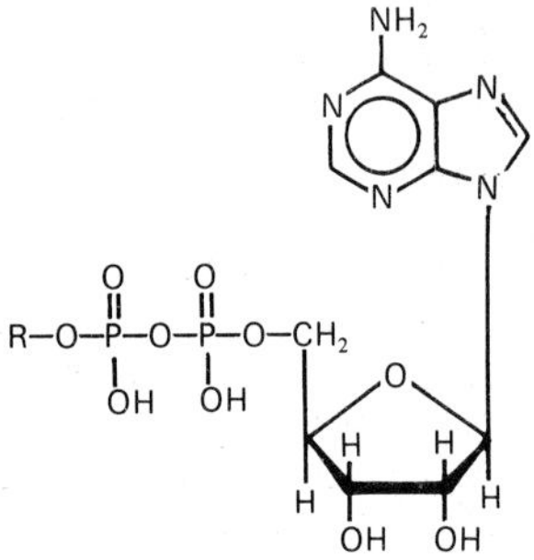

NH_2 N N N N R–O–P(=O)(OH)–O–P(=O)(OH)–O–CH_2 O H H H H OH OH

7.7

attached, but in most biological materials it occurs predominantly in the form of two nucleotides, flavin mononucleotide, FMN [7.6], and flavin-adenine dinucleotide, FAD [7.7]. These occur both as prosthetic groups in the group of respiratory enzymes known as flavoproteins and also free, when they are referred to as coenzymes.

The distribution of riboflavin in foodstuffs is very similar to that of thiamine, in that it is found at least in small amounts in almost all biological tissues and is particularly abundant in meat (0.2 mg per 100 g) especially liver (3.0 mg per 100 g). Unlike thiamine, cereals are not a particular rich source but milk (0.15 mg per 100 g) and cheese (0.5 mg per 100 g) are valuable sources. Dried brewer's yeast and yeast extracts contain large amounts of riboflavin and many other vitamins but it is interesting that the thiamine does not leak out of the yeast into the beer during fermentation but the riboflavin does (to give 0.05 mg per 100 g).

The flavoproteins of which riboflavin forms part of the prosthetic group are all involved in oxidation/reduction reactions. It is the isoalloxazine nucleus that receives electrons from substrates that are oxidised and donates electrons to substrates that are reduced:

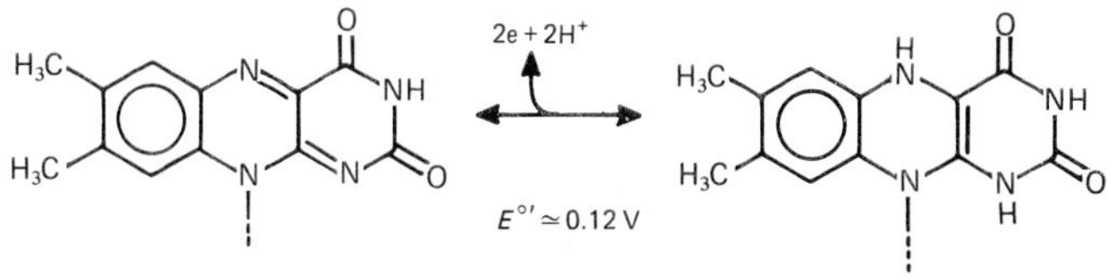

$$2e + 2H^+ \quad E^{\circ\prime} \simeq 0.12\ \text{V}$$

Most flavoprotein enzymes are involved in the complex respiratory processes that occur in the mitochondria of living cells but some are involved in other aspects of metabolism. The enzyme glucose oxidase, which was mentioned in Chapter 2 is one example. This enzyme transfers two hydrogen atoms from carbon one of the glucose to a molecule of oxygen via the FAD prosthetic group, resulting in the formation of hydrogen peroxide.

Riboflavin is one of the most stable vitamins. The alkaline conditions in which it is unstable are rarely encountered in foodstuffs. The most important aspect of riboflavin's stability is its sensitivity to light. While this is not significant in opaque foods such as meat in milk the effect can be dramatic. It has been reported that as much as 50 per cent of the riboflavin of milk contained in the usual glass bottle may be destroyed by two hours exposure to bright sunlight. The principal product of the irradiation is known as lumichrome [7.8]; at neutral and alkaline pH values some lumiflavin [7.9] is also produced. This breakdown of riboflavin has wider significance than the simple loss of vitamin activity. Both breakdown products are much stronger oxidising agents than riboflavin itself and catalyse massive destruction of ascorbic acid (vitamin C). Even a small drop in the riboflavin content of the milk can lead to a near total elimination of the ascorbic acid content. The other important side effect of riboflavin breakdown is that in the course of the reaction, which involves interaction with oxygen, highly reactive states of the oxygen molecule, in particular singlet oxygen,* are produced. Although these have only a transient existence they are able to initiate the autoxidation of unsaturated fatty acids in the milk fat, as discussed in Chapter 3. The result is an unpleasant off-flavour. When milk is to be sold in supermarkets whose chilled cabinets are often brightly lit with fluorescent lamps it is obviously essential that opaque containers rather than the traditional glass bottles are used.

---

*In singlet oxygen, written $^1O_2$, both π-electrons are paired with opposite spin in a single orbital compared with ground state or triplet oxygen, $^3O_2$, where they have parallel spin in different orbitals.

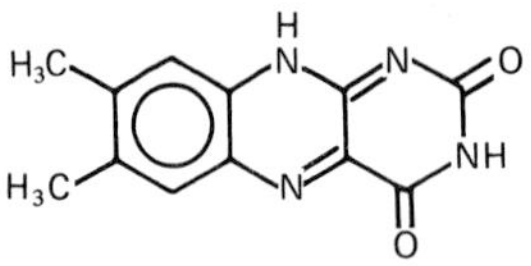

7.8

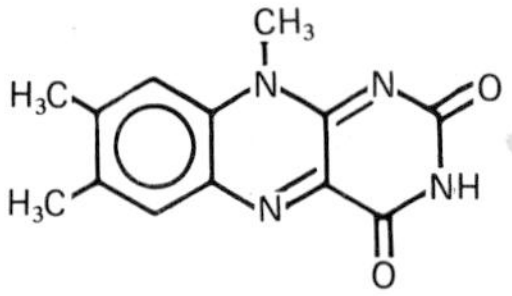

7.9

As vegetables are not generally important as sources of riboflavin losses by leaching during blanching or boiling in water are not significant. Other cooking operations including meat roasting and cereal baking have negligible effects.

Riboflavin is not readily determined in the laboratory. The most commonly applied chemical technique requires an initial treatment with 0.2M HCl at high temperatures to liberate the riboflavin from the proteins to which it is normally bound. After clean-up procedures the extract is examined fluorimetrically. The absorbance by riboflavin at 440 nm is accompanied by an emission at 525 nm.

Laboratories that routinely determine the levels of many different vitamins in food frequently use microbiological assays. These assays depend on the existence of special strains of bacteria that have lost the capability to synthesise one particular vitamin for themselves. These strains are obtained by subjecting more normal (*ie* wild type) members of the species to irradiation or mutagenic chemicals and isolating mutants that have the required characteristics. Specially prepared growth media are available which provide all the nutrients the test strain requires except the vitamin to be determined. The amount of bacterial growth in a culture using this medium will then depend on how much of the vitamin is added – either from a food sample extract or from a standard solution used for comparison. Unfortunately microbiological assays require technical skills not normally associated with chemical laboratories and they have therefore become the province of specialised establishments. However, their sensitivity, in experienced hands, is very impressive. The assay for riboflavin using a strain of *Lactobacillus casei* covers the range 0–200 ng per sample but this is one of the least sensitive. The vitamin B_{12} assay using *Lactobacillus leichmannii* has an assay range of 0–0.2 ng per sample!

Pyridoxine (vitamin B_6, pyridoxol)

The form of this vitamin that is active in the tissues, again as the prosthetic group of a number of enzymes, is pyridoxal phosphate [7.10]. However, when the vitamin is identified in foodstuffs it is in one of three forms, all having lost the phosphate group, pyridoxine (also more correctly pyridoxol) [7.11], pyridoxal, and pyridoxamine [7.12]. In foodstuffs of plant origin the first two of these usually predominate whereas in animal materials it is the last two that predominate. At least two-thirds of the vitamin in living tissues is probably present as enzyme prosthetic groups, *ie* tightly bound to protein. In muscle tissue one particular enzyme, phosphorylase, has a major share of the total. Milk is one exception, only about 10 per cent being protein bound.

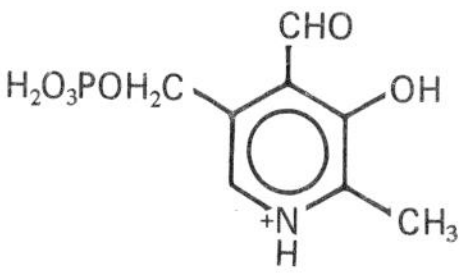

$H_2O_3POH_2C$, CHO, OH, CH_3, $\overset{+}{N}H$ — 7.10

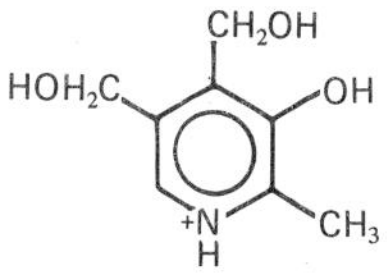

HOH_2C, CH_2OH, OH, CH_3, $\overset{+}{N}H$ — 7.11

HOH_2C, CH_2NH_2, OH, CH_3, $\overset{+}{N}H$ — 7.12

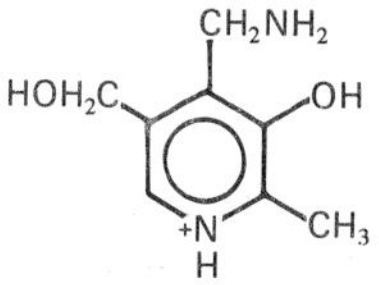

Pyridoxine (this name is used to describe all the active forms of the vitamin besides pyridoxol) is considered to be widely distributed amongst foodstuffs but the problems posed by its analysis are such that really reliable data is scarce. It has been found in at least small amounts (around 10 μg per 100 g) in almost all biological materials in which it has been sought. Meat and other animal tissues, egg yolk and wheat germ are particularly rich sources (around 500 μg per 100 g); milk and cheese have about 50 μg per 100 g.

Compared with other vitamins/prosthetic groups, pyridoxal phosphate is quite versatile in terms of the types of reactions in which it is involved. Almost all the enzyme-catalysed reactions in which it participates have amino acids as substrates, phosphorylase is one exception. The reaction sequence of a transamination, a typical pyridoxal phosphate requiring reaction, is shown in *Fig. 7.2*.

The question of the stability of pyridoxine during food processing is complicated by tendency of the different forms to differ in stability and also to interconvert during some processing operations. Pyridoxol is very stable to heat within the pH range usually encountered in foodstuffs. The other two forms are a little less stable but as with the other water soluble vitamins leaching is the major cause of losses during cooking and processing. Although there is little loss of vitamin activity during milk processing the more extreme methods, such as the production of evaporated milk, cause extensive conversion of pyridoxal, the predominant form in raw milk, to pyridoxamine. An identical change occurs during the boiling of ham. When milk is dried, extensive losses of pyridoxal can occur due to interactions with sulphydryl groups of proteins. Interaction with free amino groups of proteins leads to pyridoxamine formation – which does not cause loss of vitamin activity.

The analysis of pyridoxine levels in foods is not practical using chemical analytical techniques and therefore microbiological methods must be used. A special source of difficulty with microbiological assays

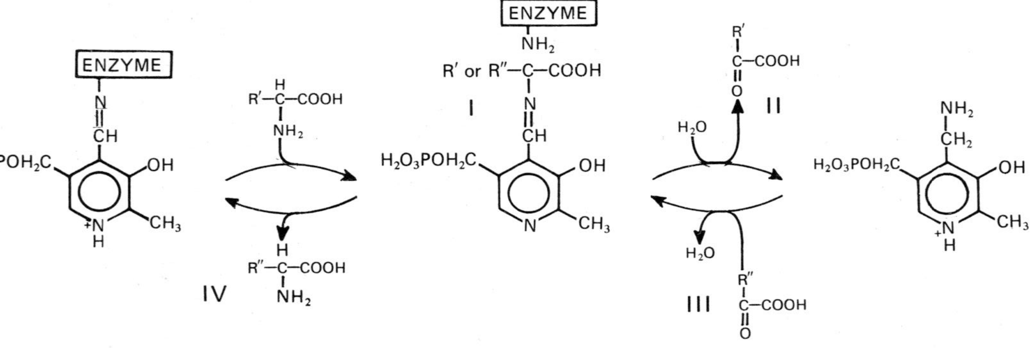

FIG. 7.2. The mechanism of transamination reactions. The ϵ-amino group of a lysine in the active site of the enzyme carries the pyridoxal phosphate. This is displaced by the incoming amino acid to form a Schiff base (I). The hydrolysis of the weakened N–C bond then liberates the α-keto acid (II). The sequence is then reversed, as the second α-keto acid (III) enters the second amino acid (IV) is formed. The result is an overall reaction such as:

glutamate + oxaloacetate $\rightleftharpoons$ α-ketoglutarate + aspartate

for this vitamin is that the different forms of the vitamin differ in potency for micro-organisms although apparently they do not as far as animal nutrition is concerned.

Niacin (nicotinic acid and nicotinamide)

Niacin is the collective name given to nicotinic acid [7.13] and its amide, nicotinamide [7.14]. In living systems the pyridine ring occurs as a component of nicotinamide adenine dinucleotide, NAD* [7.15], and its phosphate derivative, NADP* [7.16]. NAD does occur as a prosthetic group in a few enzymes such as glyceraldehyde-3-phosphate dehydrogenase but most NAD and NADP in living cells functions as an electron carrier in respiratory systems.

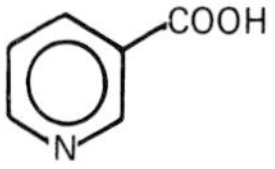

7.13 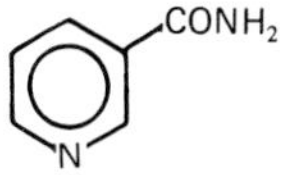 7.14

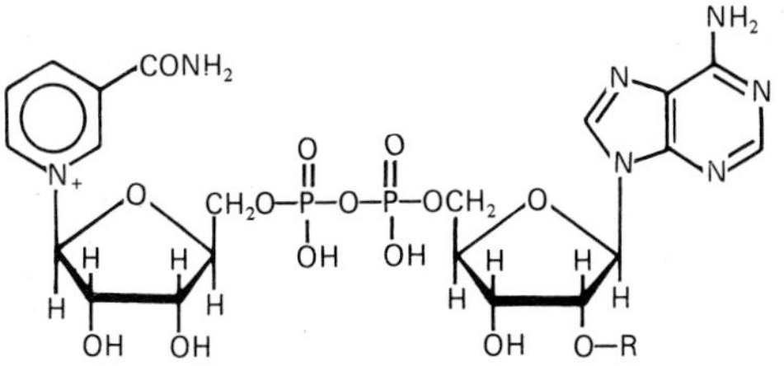

$R = H$ 7.15

$R = PO_3H_2$ 7.16

For example in most of the Krebs cycle reactions (the key metabolic pathway of respiration) the electrons removed from the organic acid substrates are transferred initially to NAD. From NAD these are then used to reduce oxygen to water in the oxidative phosphorylation system of the mitochondrion. The route from NAD involves first flavoproteins and then porphyrin ring containing proteins called cytochromes. The most striking feature of the reduction of NAD:

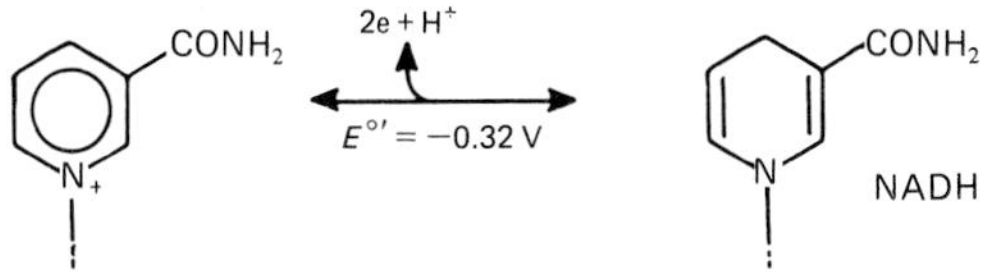

$2e + H^+$; $E^{\circ\prime} = -0.32$ V ; NADH

*In older literature these were referred to as di- and triphosphopyridine nucleotide, DPN and TPN, respectively.

is the very high absorbance of NADH at 340 nm, wheie the oxidised form has almost no absorbance at all.

In general the distribution of niacin in foodstuffs is similar to that of the other vitamins that have been discussed so far in this chapter. Although present in small amounts in all biological materials it is meat (with 5–15 mg per 100 g) that is the richest source in most diets. Dairy products, milk, cheese and eggs, are surprisingly poor sources of this vitamin having no more than 0.1 mg per 100 g). Fruit and vegetables, particularly legume seeds, are quite useful sources (0.7–2.0 mg per 100 g). Although many cereals have apparently high niacin contents (*eg* wholewheat flour 5 mg and brown rice 4.7 mg per 100 g) milling can have drastic effects as a very large proportion of the vitamin is located in the germ. To compound the problem much of the niacin in cereals is tightly bound to some of the hemicelluloses present and is not liberated by the digestive processes of the alimentary tract. This is a special problem with maize. Pellegra, the disease caused by a deficiency of niacin, was commonly associated with maize being a staple foodstuff. Pellegra was largely unknown in Europe until maize became a major crop in the Mediterranean region during the 18th century and it was also common in the undernourished communities of the southern states of the US. However, in the one country where pellegra might be expected to be widespread, Mexico, it was not. The reason is that in Mexico *tortillas*, which are the dietary equivalent of bread, are traditionally made from maize flour which has been treated with lime water. We now know that under alkaline conditions bound, unavailable niacin is liberated by heating!

Niacin is stable under most cooking and processing conditions, leaching is the only significant cause of losses. During the baking of cereals, especially in products made slightly alkaline by the use of baking powder the availability of niacin may actually increase.

The usual method for the determination of niacin is a microbiological assay.

Cobalamin (cyanocobalamin, vitamin B_{12})

The chemical structure of this vitamin is by far the most complex of all the vitamins. As shown in *Fig. 7.3*, its most prominent feature is a 'corrin' ring which resembles a porphyrin ring system but differs in two respects. The ring substituents are mostly amides and the coordinated metal atom is cobalt. The fifth coordination position is occupied by a nucleotide-like structure based on 5,6-dimethylbenzimidazole which is also attached to one of the corrin ring amide groups. The sixth co-ordination position (R in *Fig. 7.3*), on the opposite face of the corrin from the fifth, can be occupied by a number of different groups. As normally isolated the vitamin contains a —CN group in the sixth position – hence the name cyanocobalamin. In its functional state, when it is the prosthetic group of a number of enzymes it may have either a methyl group or a 5′-deoxyadenosyl residue in the sixth position. The coordination of the methyl and 5′-deoxyadenosyl residues is unusual in that it is a carbon atom (the 5′ in the case 5′-deoxyadenosyl) that is linked to the cobalt atom.

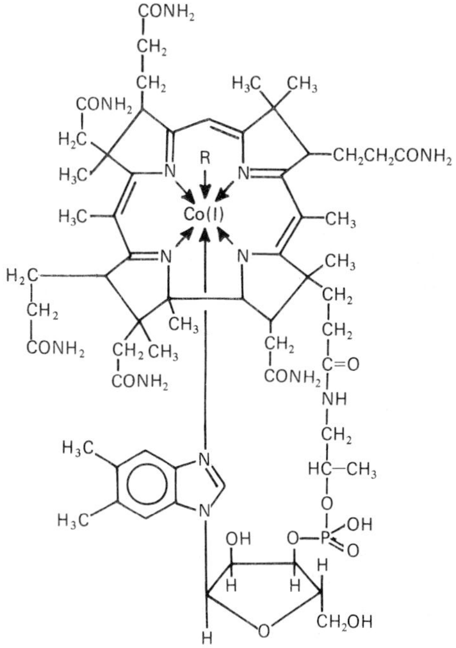

FIG. 7.3. The structure of cobalamin.

The distribution of cobalamin in foodstuffs is most unusual. Bacteria are the only organisms capable of its synthesis so that it is never found in higher plants. When traces have been detected on vegetables the source is almost certainly surface contamination with faecal material that has been used as fertiliser. The bacterial flora of the colon secretes about 5 μg daily but since there is no absorption from the colon this is wastefully excreted. The daily requirement is about 1 μg. The vitamin occurs in useful quantities in most foodstuffs of animal origin (per 100 g: eggs 0.7 μg, milk 0.3 μg, meat (muscle tissue) 1-2 μg, liver 40 μg, kidney 20 μg) but in all these cases the vitamin got there either from micro-organisms or animal tissues consumed in the diet or from synthesis by intestinal micro-organisms high enough in the alimentary tract for absorption to occur.

As a result of the low requirement for this vitamin and its abundance in many foodstuffs simple deficiency is extremely rare except amongst dietary extremists such as vegans. It has even been suggested that the slowness of the onset of deficiency symptoms in those young children unfortunate enough to have vegan parents is caused by the lack of toilet hygiene that is inevitable in young children providing a source of the vitamin from unwashed hands. The classical disease of apparent cobalamin deficiency is pernicious anaemia. This is caused by a failure to absorb the vitamin rather than a lack of it in the diet. Patients with pernicious anaemia are unable to synthesise a carbohydrate rich protein known as the *intrinsic factor*. This is normally secreted by the gastric mucosa and complexes the vitamin, facilitating its absorption in the small intestine.

The catalytic activities of cobalamin enzymes are almost all associated with one-carbon units such as methyl groups, an area of metabolism shared in bacteria and animals with folic acid containing enzymes and taken over by them in plants. The role of cobalamin enzymes in methane synthesis by bacteria has a superficial similarity to that of Grignard reagents. An important example of the synthetic dexterity of cobalamin-containing enzymes is the isomerisation of methylmalonyl coenzyme A to succinyl coenzyme A:

$$\begin{array}{c} O{=}C{-}S{-}CoA \\ | \\ H{-}C{-}CH_3 \\ | \\ COOH \end{array} \longrightarrow \begin{array}{c} O{=}C{-}S{-}CoA \\ | \\ CH_2 \\ | \\ CH_2 \\ | \\ COOH \end{array}$$

The relationship between the known biochemical functions of cobalamin and the clinical manifestations of its deficiency, a drastically reduced synthesis of the haem moiety of haemoglobin, remains obscure.

In spite of the complexity of its structure cobalamin is fairly stable to food processing and cooking conditions and what losses do occur are not sufficient to cause concern to nutritionists. As with all the water soluble vitamins, leaching is the major cause of loss. Microbiological assay is the only practical method of determining the vitamin in foodstuffs.

Folic acid (folacin)

Folic acid is the name given to a closely related group of widely distributed compounds whose vitamin activity is similar to the cobalamins. The structure of folic acid proper is shown in *Fig. 7.4*.

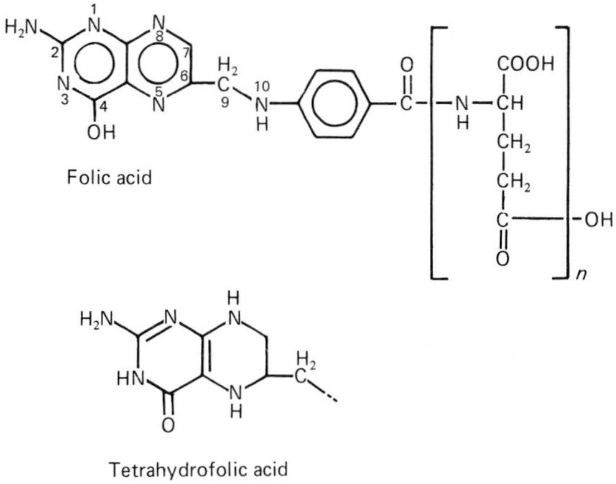

Folic acid

Tetrahydrofolic acid

FIG. 7.4. The structure of folic acid and tetrahydrofolic acid.

The form that is active as a coenzyme is the tetrahydrofolate with the pteridine nucleus of the molecule in its reduced state and several (*n* between 3 and 7) glutamate residues carried on the *p*-aminobenzoate moiety. Accurate determinations of the folic acid content of foodstuffs are difficult, even using microbiological assays, as it is necessary to remove all but one of the glutamyl residues before consistent results are obtained. This can now be achieved by enzymic methods but it has meant that many earlier estimates of the folic acid content of foodstuffs were orders of magnitude too low. Only recently have measurements in terms of particular glutamyl/folic acid conjugates been made. Of animal tissues, where the pentaglutamyl conjugate predominates, liver is a particularly good source (300 μg per 100 g); muscle tissues have much less (3–8 μg per 100 g). Leafy green vegetables, where the heptaglutamyl conjugate predominates, have around 20–80 μg per 100 g and probably constitute the most important dietary source.

As a coenzyme tetrahydrofolate (FH_4) is concerned solely with the metabolism of one-carbon units, which it carries at N-5 or N-10. While bound to the coenzyme the one-carbon unit may be oxidised or reduced and undergo other changes giving the structures listed in *Fig. 7.5*. These are intimately involved in the metabolism of many amino acids, purines and pyrimidines (components of nucleotides and nucleic acids) and, indirectly, haem. In animals the metabolism of methyl groups, which must be supplied in the diet (mostly by choline), is particularly complex. Both folic acid- and cobalamin-requiring reactions are involved so that it is not surprising that there is some overlap in the symptoms of deficiencies of these two vitamins. There is growing support for the view that deficiency of folic acid is widespread, even in the comparatively well fed populations of Europe and the US.

Although it is now common to supply folic acid supplements to pregnant women there is opposition to routine dietary fortification. This is because extra folic acid will alleviate, and therefore mask, the obvious anaemia of cobalamin deficiency. However, it has no effect on the accompanying neurological damage that cobalamin deficiency causes which, by the time it is detected directly, cannot be reversed.

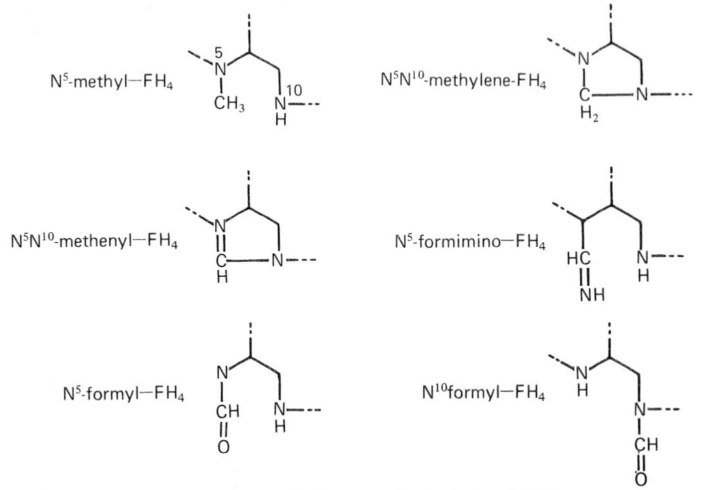

N^5-methyl–FH_4 N^5N^{10}-methylene-FH_4

N^5N^{10}-methenyl–FH_4 N^5-formimino–FH_4

N^5-formyl–FH_4 N^{10}formyl–FH_4

FIG. 7.5. One-carbon units carried by tetrahydrofolate (FH_4).

There is very little useful information on the stability of folic acid and its derivatives. The stability to heat and acidity varies with both the number of glutamyl residues and the nature of any attached one-carbon units. Oxidation can be a problem but ascorbic acid has a valuable protective effect. As with the other water soluble vitamins most cooking and processing losses are caused by leaching.

The fact that animals depend on their diet for their supply of folic acid makes possible the use of sulphonamides such as sulphanilamide [7.17] as antibacterial drugs. These substances interfere with the formation of the folic acid that the bacteria themselves require without having any great effect on folic acid metabolism in the patient.

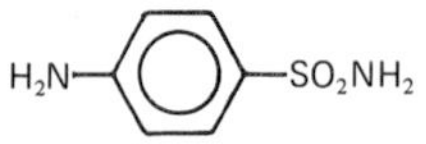

H_2N–C_6H_4–SO_2NH_2

7.17

Biotin and pantothenic acid

These two vitamins have little in common in terms of their structure or function. What brings them together is the total absence of reports of deficiency symptoms in man or animals except under the artificial conditions of the laboratory. Biotin [7.18] occurs as the prosthetic group of many enzymes catalysing carboxylation reactions such as the conversion of pyruvate to oxaloacetate. Most foodstuffs contain at least a few micrograms per 100 g (assayed microbiologically) and there appears to be a more than adequate amount in any otherwise satisfactory diet. The only known deficiency of biotin occurs when extremely large quantities of raw eggs are eaten. Egg white contains a protein, avidin, which complexes biotin. Avidin is denatured when eggs are cooked. Biotin is apparently quite stable to cooking and processing procedures.

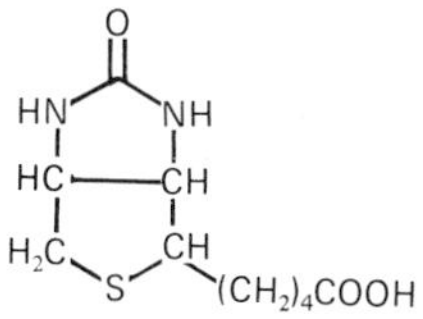

7.18

As with biotin much more is known about the biochemical function of pantothenic acid (*Fig. 7.6*) than about its status as a food component. It occurs in two forms in living systems, as part of the prosthetic group of the acyl carrier protein (ACP) component of the fatty acid synthetase complex and as part of coenzyme A, the carrier of acetyl (ethanoyl) and other acyl groups in most metabolic systems. In both of these roles the acyl group is attached by a thioester link (see *Fig. 7.6*) to the sulphydryl group which replaces the carboxyl group

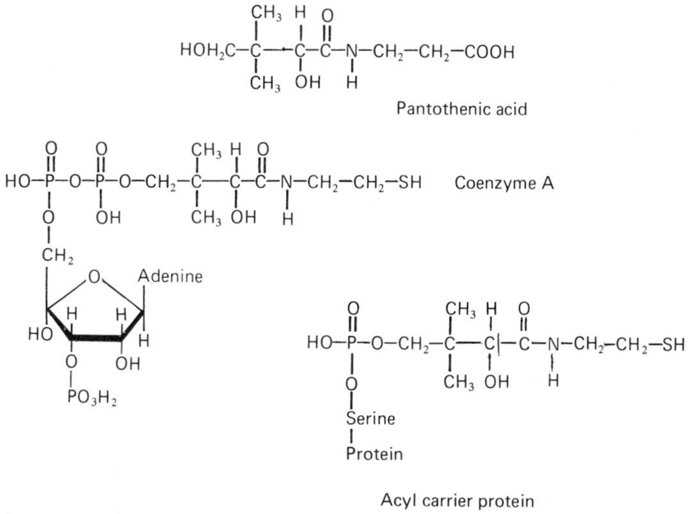

Pantothenic acid

Coenzyme A

Acyl carrier protein

FIG. 7.6. Pantothenic acid and its functional derivatives.

of the free vitamin. Pantothenic acid is extremely widely distributed, mostly at around 0.2–0.5 mg per 100 g, levels generally assumed to give a more than adequate dietary intake. As with so many other vitamins liver contains especially high levels, in this case about 20 mg per 100 g. Quite high losses of pantothenic acid have been reported to occur during food processing mostly due to leaching but some decomposition is known to occur especially during heating under alkaline or acid conditions.

A number of other substances including *meso*-inositol, *p*-aminobenzoic acid, lipoic acid and choline are sometimes listed with the vitamins. There is no evidence to suggest that this listing is justified as far as man is concerned.

Ascorbic acid (vitamin C)

Ascorbic acid is a rather curious vitamin. It occurs widely in plant tissues (where its function is unknown) and it is also synthesised, and is not therefore for them a vitamin, by all mammals except the primates, the guinea pig and some fruit eating bats. This latter group's only discernible common feature is a liking for fruit, a rich source of the vitamin. The structure of L-ascorbate (pK_a is 4.04 at 25°) and some related substances are shown in *Fig. 7.7*. The oxidised form, dehydro-L-ascorbic acid (an unfortunate, if inevitable, name, since the molecule has no ionizable groups) is distributed with the reduced form in small, rarely accurately determined proportions. Dehydro-L-ascorbic acid has virtually the same vitamin activity as L-ascorbate. The mirror image isomer, most commonly referred to as erythorbic acid, has no vitamin activity but does behave similarly to ascorbate in many food redox systems and therefore finds some similar applications as a food additive.

Much more is known about the distribution of ascorbic acid in foodstuffs than about any other vitamin. The reason is not that ascorbic

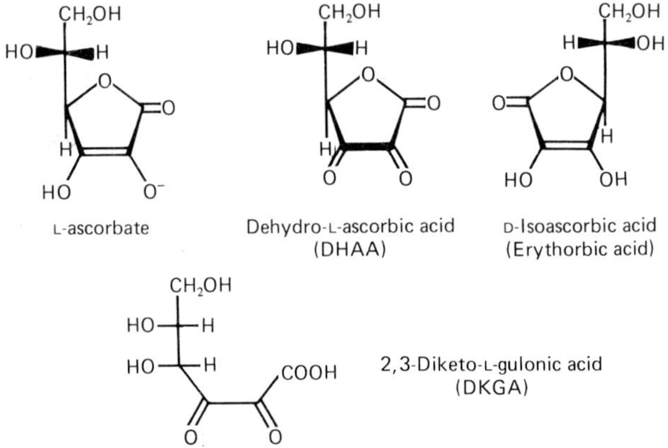

FIG. 7.7. The structure of ascorbic acid and related substances.

acid is inherently more interesting or more important but simply because it is much easier to determine than any of the others so that when a research worker or student wishes to do a project 'on vitamins' this is the one that usually gets chosen. The richest sources are fruits but, as the data in Table 7.1 show, wide variations do occur. It has to be remembered that a foodstuff that is eaten in fairly large quantities will be an important dietary source even if it contains only modest levels of the vitamin. The very high levels to be found in the parsley are nutritionally insignificant when compared with the quantity in the new potatoes it is used to garnish. Cereals contain only traces and of the animal tissue foods only liver (30 mg per 100 g) is a valuable source. Cow's milk, and other dairy products contain insignificant amounts of the vitamin.

Table 7.1. Typical ascorbic acid contents of some fruit and vegetables (mg/100 g).

Plum	3	Carrot	6
Apples	6	Peas	25
Peach	7	New potatoes	30
Pineapple	25	Old potatoes	8
Tomato	25	Cabbage	40
Citrus fruit	50	Cauliflower	60
Strawberry	60	Broccoli	110
Blackcurrant	200	Horseradish	120
Rose hips	1000	Parsley	150

The content of the edible part when fresh and unprocessed is shown.

The role of ascorbic acid in mammalian physiology and biochemistry is far from fully understood. It has been implicated in a number of reactions, particularly in the formation of hydroxyproline and hydroxylysine from the parent amino acids *after* their incorporation into the polypeptide chain of collagen. Most of the symptoms of scurvy can be related to a failure of normal collagen biosynthesis. For example, wounds fail to heal, internal bleeding occurs and the joints become painful. Another important reaction requiring ascorbate is the hydroxylation of 3,4-dihydroxyphenylethylamine (dopamine, a derivative of tyrosine) to noradrenaline:

$$\text{(3,4-(OH)}_2\text{C}_6\text{H}_3\text{)}-CH_2-CH_2-NH_2 \xrightarrow[\text{DHAA}]{\text{ascorbate}} \text{(3,4-(OH)}_2\text{C}_6\text{H}_3\text{)}-\text{C(H)(OH)}-CH_2-NH_2$$

This may be the origin of the disturbances of the nervous system that accompany scurvy. There is also increasing evidence that ascorbate has a number of roles in the absorption of iron from the intestine and its subsequent transport in the bloodstream although little is firmly established. It is perhaps this lack of firm information on ascorbic acid's activities that has facilitated its adoption as a 'wonder cure-all' for numerous other conditions and diseases, particularly the common cold. The actual evidence for the value of extra-high doses (up to 5 g per day *ie* over 100 times the recommended dietary intake) is extremely flimsy, something that cannot be said for the profits to be made in selling ascorbic acid to 'health food' enthusiasts.

There are many different routes available for the chemical synthesis of ascorbic acid but the most usual commercially adopted route is shown in *Fig. 7.8*.

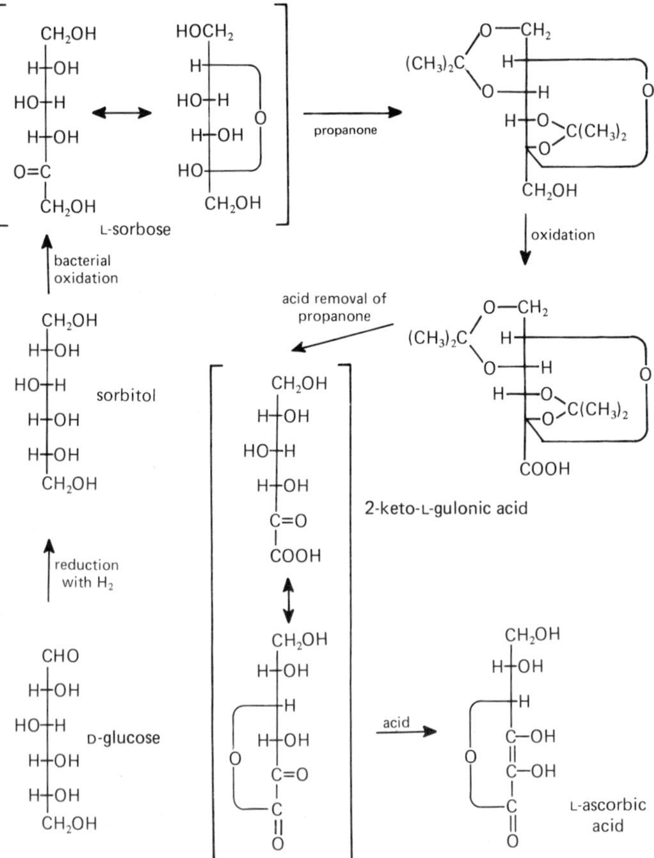

FIG. 7.8. The chemical synthesis of ascorbic acid from glucose.

Almost all of the numerous methods that have been devised for the determination of ascorbic acid depend on measurements of the reducing property by titration with an oxidising agent, the most popular one being 2,6-dichlorophenolindophenol (DCPIP) [7.19]. DCPIP is blue in its oxidised form, colourless when reduced. Highly coloured plant extracts such as blackcurrant juice obviously present special problems as the endpoint of the titration is obscured. There are numerous, more or less satisfactory, techniques for circumventing the problem. One example is titration with *N*-bromosuccinimide in the presence of potassium iodide and ether. Excess *N*-bromosuccinimide liberates I_2 from the iodide and this dissolves in the ether phase to give it a brown colour when the endpoint is reached. Very few of the methods commonly used determine any DHAA present; one that does not discriminate between the two compounds is the fluorimetric method using *o*-phenylenediamine.

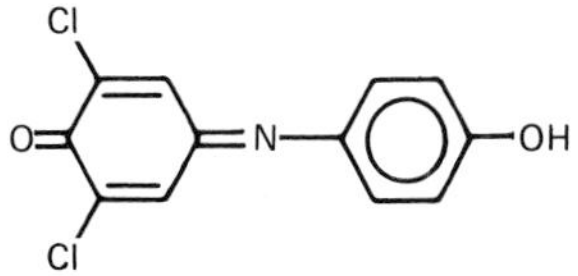

Cl

O= ... =N— ... —OH

Cl

7.19

The information that has been gained by these studies shows that ascorbic acid is one of the least stable vitamins. Losses during processing, cooking or storage can occur by a number of different routes but unfortunately very few experiments have been devoted to discovering which routes are involved in particular foodstuffs. Leaching is obviously important in vegetable preparation when the cooling or processing water is discarded. Green vegetables may lose more than 50 per cent of the vitamin in this way if they are boiled for prolonged periods. Losses from root vegetables are usually much smaller simply because they present a smaller surface area in relation to their weight.

In the presence of air most actual degradation of the vitamin occurs through the formation of the less stable DHAA. Once formed DHAA rapidly undergoes an irreversible ring opening reaction to form DKGA, which has no vitamin activity. The oxidation to DHAA can occur by various mechanisms. In plant tissues stored without blanching, especially if they have been sliced, peeled or otherwise damaged, the enzyme ascorbic acid oxidase will be very active, catalysing the reaction:

$$\text{ascorbate} + O_2 \rightarrow \text{DHAA} + H_2O_2$$

The enzyme phenolase is also responsible for ascorbic acid losses when, as was discussed in Chapter 5, ascorbic acid reduces *o*-quinones back to the original *o*-diphenols. Metal cations, particularly Fe^{3+} and

Cu^{2+}, will catalyse the oxidation and may cause serious losses in food products. Even in the absence of a catalyst the oxidation occurs quite rapidly at elevated temperatures. Formation of DKGA is virtually instantaneous at alkaline pH values, rapid around neutrality and slow under acid conditions. The reason for the undesirability of adding sodium bicarbonate to green vegetable cooking water to preserve the colour now becomes clear.

In fruit juice processing the acidity ensures that the vitamin is quite stable. The stability is enhanced by the presence of citrate and flavanoids which both complex metal cations but it is still necessary to keep juices de-aerated as much as possible. At elevated temperatures under anaerobic conditions ascorbic acid will also undergo breakdown along similar lines to other sugars (see Chapter 2) except that CO_2 is evolved. However, other details of the pathway remain the subject of speculation.

One other aspect of ascorbic acid in foodstuffs remains to be pointed out. Its cheapness and its obvious acceptability as a nutrient make it a valued food additive for technological rather than nutritional purposes. In fruit and vegetable products, notably dehydrated potato, it is used as an antioxidant to prevent the browning reactions that would be catalysed by phenolase. In cured meat products it is used as a reducing agent to lower the concentration of nitrite needed for a good pink colour and in modern bakeries it is used as a flour improver. All these applications are considered in greater detail in the appropriate chapters of this book.

Retinol (vitamin A)

This vitamin is the first of the 'fat-soluble' vitamins to be considered in this chapter. The form that is active in mammalian tissues is the alcohol retinol [7.20] but in the diet most of the vitamin occurs as precursors, of which there are several.

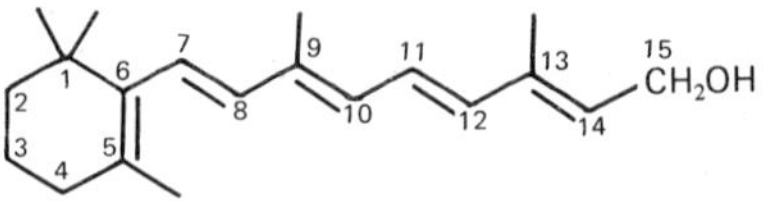

7.20

Plant tissues provide the vitamin in the form of carotenoids. As was indicated in Chapter 5 any carotenoid that contains the β-ionone ring system, as do β-carotene (at both ends of the molecule), α- and γ-carotene and β-apo-8′-carotenal (at one end only), is converted by enzymes in the mucosa of the small intestine to retinol. The absorption of carotenoids and their conversion to retinol does not occur with total or uniform efficiency so that estimation of the vitamin activity of different foodstuffs is rather complicated. As a general rule 6μg of β-carotene or 12μg of other *active* carotenoids is regarded as the equivalent of 1μg of retinol. Leafy green vegetables contain between 1000 and 3000 retinol equivalents (*ie* an amount equivalent to 1μg of retinol) per 100 g, mostly as β-carotene. Carrots, the classic source, contain about 2000 retinol equivalents per 100 g. The only other major plant source is red palm oil whose β-carotene content is sufficient to give some 20 000 retinol equivalents per 100 g.

In animal tissues the vitamin (always originally derived from plant sources) is stored and transported as the retinyl ester of long chain fatty acids, mostly palmitate and stearate. Dietary retinyl esters are first hydrolysed and then re-esterified by the intestinal mucosal cells during the process of absorption. A specific protein binds the retinyl esters in the liver so that liver and liver oils are particularly good sources in the diet. Cod liver oil and halibut liver oil are well known for their vitamin contents, around 10^5 and 10^7 retinol equivalents per 100 g respectively. The vitamin content of mammalian liver varies widely, from 3000 in pig's liver and 17 000 in sheep's liver to 600 000 retinol equivalents per 100 g in polar bear liver! The relationship between the diet of these animals and the vitamin contents of their livers is obvious. Meat contains very little of this vitamin. Although milk itself has only a low concentration of the vitamin dairy products where the lipid phase of the milk is concentrated, such as butter and cheese, contain 500–1000 retinol equivalents per 100 g. Margarine is supplemented with synthetic retinyl acetate or sometimes red palm oil to give it a similar vitamin activity to butter.

The physiological roles of retinol are mostly poorly understood. It is known that specific enzymes in the eye convert retinyl esters to retinol which associates with the proteins known as *opsins*. The resulting complexes are the visual pigments of the retina. The response of these proteins to light involves an interconversion of retinal and its 11-*cis* isomer. Night blindness was linked to a deficiency of this vitamin as early as 1925. We know much less about the other functions of retinol. Deficiency of the vitamin leads ultimately to death due to effects quite unconnected with blindness. These include *xerophthalmia*, a drying and degenerative condition of the cornea of the eye which, untreated, will cause total blindness, abnormal bone development, and disorders of the reproductive system. Although the nature of retinol's function in these tissues has not been resolved there are indications of an involvement in biosynthesis of some glycoproteins that occur in mucous membranes.

An unusual feature of retinol is its toxicity in excess. Most victims of this hypervitaminosis have been 'health food' enthusiasts but it has been reported that Polar explorers have also suffered following the ill-advised consumption of a polar bear. As the recommended daily intake for the vitamin is only about 750 retinol equivalents one modest portion of the liver is equivalent to over two year's supply of retinol.

Both retinyl esters and β-carotene are fairly stable in food products. Most information on stability concerns β-carotene. High temperatures in the absence of air, as in for example, canning, can cause isomerisation to *neo*-carotenes (see Chapter 5, p 111) which only have vitamin activity if one end of the molecule remains unaffected. In the presence of oxygen, breakdown can be rapid, especially in dehydrated foods where a large surface area is exposed to the atmosphere. The hydroperoxides that result from the autoxidation reactions of polyunsaturated fatty acids will bleach carotenoids and can be presumed to have a similar destructive effect on retinyl esters.

The determination of the amount of this vitamin in a foodstuff is complicated by the diversity of forms in which it occurs. An indication can be gained from the absorbance of a solvent extract at 328 nm, the absorption maximum of retinol. Elaborate procedures are available to compensate for the interfering absorbance of other substances at this wavelength but nowadays this problem is usually avoided by preliminary column chromatography to separate and identify the different carotenoids and retinyl esters.

Cholecalciferol (vitamin D, calciferol)

Only one form of this vitamin, known either as cholecalciferol [7.21] or vitamin D_3 is usually encountered in the diet. Ergocalciferol [7.22] (vitamin D_2) has vitamin activity but since it is only encountered as the synthetic derivative of a plant sterol it nowadays has little dietary significance.* There is an increasing tendency for physiologists, but not nutritionists, to regard cholecalciferol as a hormone rather than a vitamin. This is because, as we shall see, its function in the body is that of a hormone rather than as an enzyme prosthetic group like most

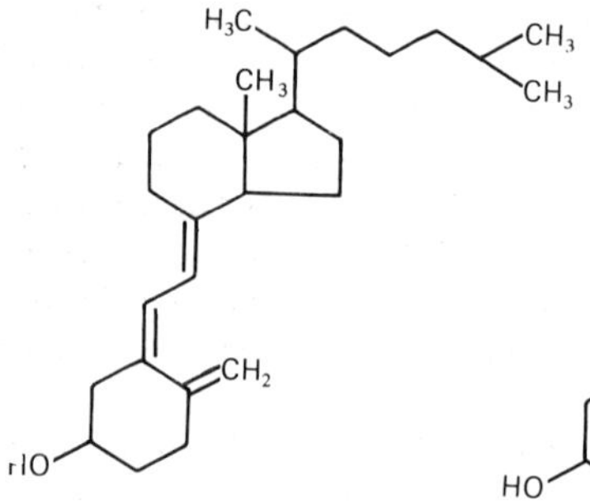

H_3C CH_3 CH_3 CH_3 CH_2 HO

7.21

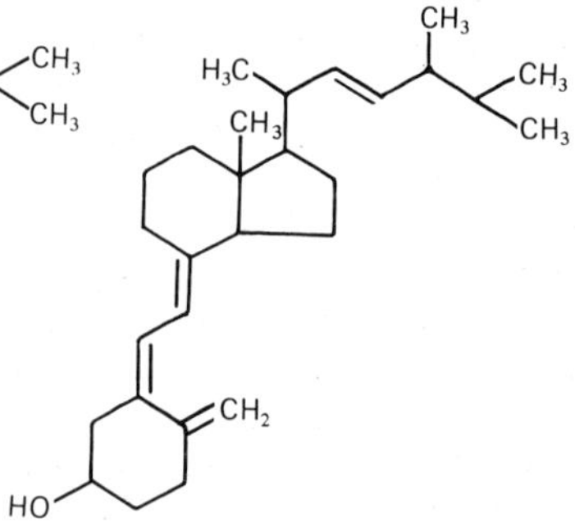

CH_3 H_3C CH_3 CH_3 CH_3 CH_2 HO

7.22

*Some of the first vitamin D supplements were based on ergocalciferol.

other vitamins. Furthermore, man and other mammals have, under the right circumstances, more than adequate capacity for synthesising it for themselves.

Animals are able to synthesise cholesterol and other steroids for themselves and one intermediate in the pathway is 7-dehydrocholesterol. The epidermal cells of the skin contain 7-dehydrocholesterol and this is converted to cholecalciferol by the action of the uv component of sunlight. The suggested course of the reaction is shown in *Fig. 7.9*.

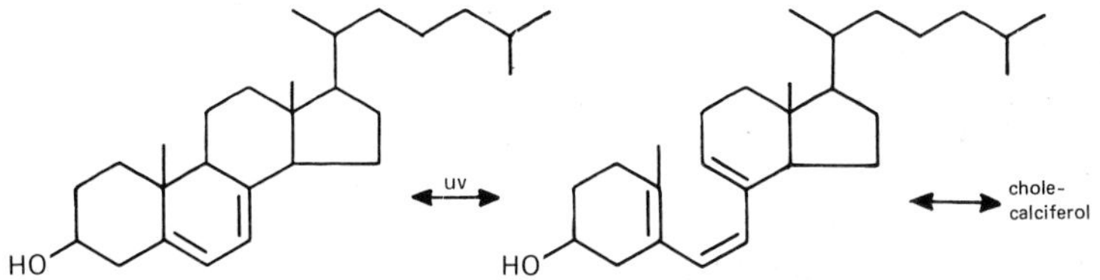

HO uv HO chole-calciferol

FIG. 7.9. The formation of cholecalciferol. The optimum wavelength for the first reaction is around 300 nm, the second reaction is spontaneous.

When human skin is adequately exposed to sunlight the body is able to provide sufficient for its needs (2.5–10 μg per day – infants, children and pregnant or lactating women have the highest needs). Storage of the vitamin in the liver allows for the variations in the supply in the diet or from irradiation of the skin. A supply in the diet becomes important in parts of the world that do not get much sunshine especially since such regions are often too cold to encourage sunbathing. It may well be that the pale complexions of the northern European races are an adaptation to ensure the most efficient use of the little sunlight available.

The pattern of distribution of cholecalciferol in foodstuffs is strikingly similar to that of retinol. As with retinol the liver oils of fish have spectacular levels, *eg* mackerel liver oil has 1.5 mg per 100 g. The muscle tissues of fatty fish such as salmon, herring or mackerel are valuable dietary sources (5–45 μg per 100 g) but mammalian tissues, including liver, have less than 1 μg per 100 g. Milk contains only about 0.1 μg per 100 ml but fatty dairy products such as butter and cream do contain useful amounts (1–2 μg per 100 g). Egg yolk has about 6 μg per 100 g. An unfortunate contrast with retinol is that plant materials contain no useful amounts of the vitamin.

It is nowadays common practice to enhance vitamin contents of certain foods including breakfast cereals, milk (in the US) and margarine. In the UK there is a legal requirement for the vitamin content of margarine to correspond to 2.0–2.5 μg of cholecalciferol per 100 g, a level similar to that found in butter.

The role of this vitamin is only now being properly elucidated. In the early years of this century it was shown that the childhood disease of rickets, a failure of proper bone development, could be prevented by a dietary component or by irradiation with uv light. There has thus been a tendency to regard rickets as a vitamin deficiency disease rather than, more correctly, as a sunlight deficiency disease. Whether from the diet or formed in the skin, cholecalciferol is converted by the body to the physiologically active compound 1,25-dihydroxycholecalciferol (calcifetriol) [7.23]. This is one of the three hormones (the others are *calcitonin* and *parathormone*) that together control calcium metabolism. Calcifetriol promotes the synthesis of the proteins that transport calcium and phosphate ions through cell membranes. A lack of calcifetriol prevents the uptake of calcium from the intestine and the resulting shortage of calcium for bone growth is manifested as rickets.

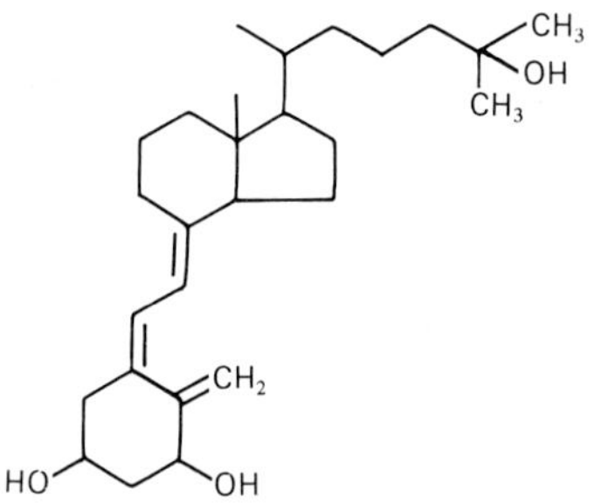

7.23

Another feature of this vitamin that it has in common with retinol is toxicity in excess. Babies and young children are the usual victims, the result of confusion in the dosages required of halibut liver oil and cod liver oil (3000 μg and 250 μg per 100 g respectively). Massive overdoses cause calcification of soft tissues such as the lungs and kidneys but in babies even modest overdoses will cause intestinal disorders, weight loss and other symptoms.

At the levels occurring in most foodstuffs the determination of cholecalciferol is extremely difficult. Biological assays using rats reared on rickets-inducing diets were essential until the recent developments of GC and HPLC* methods. As a result very little work has been done on the behaviour of cholecalciferol in food. In general it is believed to be fairly stable to heat processing and only subject to oxidative breakdown in dried foods such as breakfast cereals.

---

*HPLC – High performance liquid chromatography.

Vitamin E (α-tocopherol)

Of all the vitamins this one and also the one that concludes this chapter, vitamin K, have defeated efforts to replace their old alphabetical designations with a more scientific terminology. In both cases this is because no single chemical name is both sufficiently comprehensive and sufficiently exclusive. The vitamins E are a group of derivatives of 6-hydroxychroman carrying a phytyl side chain. As shown in *Fig. 7.10*, tocopherols have a fully saturated side chain while in tocotrienols it is unsaturated. Variations in the degree of methylation of the chroman nucleus give the α, β, γ and δ members of each series. All eight of the compounds shown in *Fig. 7.10* are found in nature but only α-, β- and γ-tocopherols and α- and β-tocotrienols are widespread.

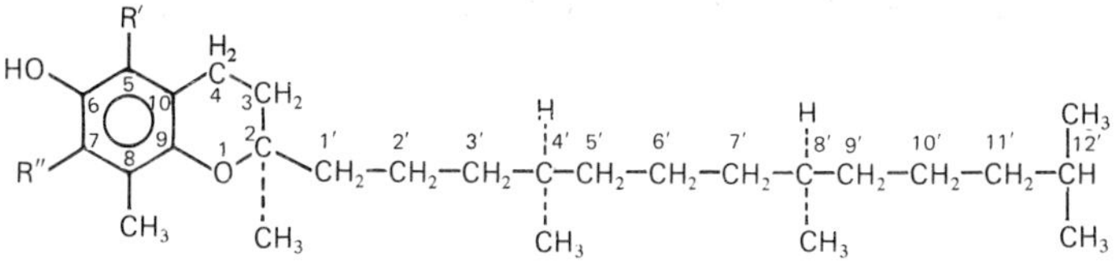

General tocopherol structure

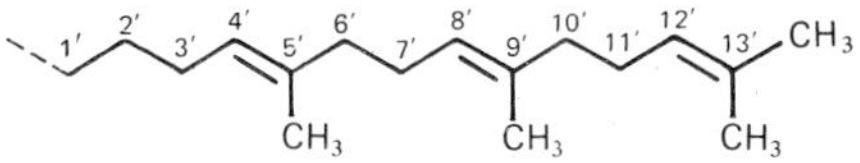

Tocotrienol side chain

	R′	R″
α	$-CH_3$	$-CH_3$
β	$-CH_3$	$-H$
γ	$-H$	$-CH_3$
δ	$-H$	$-H$

The naturally occurring tocopherols and tocotrienols are all members of the *d* series, having the configurations of their asymmetric centres (2, 4′ and 8′) as shown here. The synthetic *dl*-tocopherols are mixtures of all the eight possible optical isomers.

FIG. 7.10. Vitamin E structures.

The most important sources of vitamin E in the diet are the plant seed oils (50–200 mg per 100 g). Most other plant tissues contain less than 0.5 mg per 100 g. Animal tissues, including liver, milk and eggs have similarly low levels. Unlike other fat soluble vitamins this one is not found in particularly large amounts in fish liver oils; cod liver oil has ~25 μg per 100 g. The valuation of data such as these is complicated by the variations in vitamin activity between the different tocopherols and tocotrienols. Relative to *α-d*-tocopherol (1.0) the vitamin activities of the β-, γ- and δ- *d*-forms are 0.27, 0.13 and 0.01 respectively. The *dl* racemic mixtures have about three-quarters the activity of the corresponding pure *d* isomers. Of the tocotrienols only the α- form has significant vitamin activity (about 0.3). Table 7.2 shows the distribution of the various vitamins E in some important dietary sources. Although α-*d*-tocopherol is the most potent, these data show that the less potent but much more abundant γ-*d*-tocopherol is at least as important in nutritional terms.

Table 7.2. The relative proportions of tocopherols and tocotrienols in some seed oils.

	d-Tocopherols/per cent				Total tocotrienols/per cent
	α	β	γ	δ	
Corn oil	18	–	81	1	–
Cotton seed oil	51	–	49	–	–
Soya bean oil	11	–	66	23	–
Wheatgerm oil	60	34	–	–	6

The tocopherols are the natural antioxidants of animal (and possibly plant) tissues. Vegetable oils, such as those listed in Table 7.2, are also protected by their tocopherol content. Antioxidants block the free radical chain reactions of lipid peroxidation (see Chapter 3). The tocopherols have a special affinity for the membrane lipids of the mitochondria and endoplasmic reticulum of animal cells. It has been suggested that the shape of the tocopherol side chain will permit the formation of a complex with the arachidonic acid component of the phospholipids of these membranes. The mitochondrial membranes from animals reared on vitamin E deficient diets are exceptionally prone to peroxidation.

A deficiency of dietary selenium can cause similar symptoms in farm and laboratory animals to those invoked by vitamin E deficiency. The only known physiological role of selenium is in the active site of the enzyme glutathione peroxidase, the enzyme responsible for the breakdown of hydrogen peroxide and organic peroxides and so constituting a second defence mechanism against the consequences of lipid autoxidation.

In spite of health food enthusiasts' habit of regarding vitamin E as beneficial to all aspects of human sexual behaviour* no deficiency disease or condition has ever been identified in man under otherwise normal circumstances. It is now well established that in order to maintain what are thought to be satisfactory tissue levels of the vitamin the minimum intake is related to the quantity of polyunsaturated fatty

*The most studied symptom of vitamin E deficiency in laboratory animals is reproductive failure caused by foetal resorption or testicular degeneration.

acids in the diet. A diet high in vitamin E, rich in vegetable oils will, unfortunately, also be rich in the polyunsaturated fatty acids that these oils contain (see Chapter 3).

Determination of vitamin E in foodstuffs is complicated by the diversity of forms with different vitamin activity. Straightforward chemical tests for total tocopherols can be carried out but they are clearly valueless. Chromatographic methods, GC, TLC and HPLC, are now established which can provide a distinction between the different forms.

Little is known of the stability of vitamin E in the foodstuffs that contain only modest proportions. In extracted vegetable oils it is quite stable unless conditions allowing autoxidation of the polyunsaturated fatty acids arise. Its presence in the oil will delay the onset of rancidity in the oil but inevitably the build-up of peroxides in the oil will eventually overcome the tocopherols and cause their oxidation to compounds lacking antioxidant activity. Vitamin activity is rapidly lost from the oil content of commercially deep-fat-fried frozen products such as potato chips. Low temperature storage does not prevent the loss of most of the vitamin activity.

Vitamin K (phylloquinone, menaquinones)

This is the last of the fat-soluble vitamins to be considered. As has already been mentioned the alphabetical designation has been maintained in view of the lack of suitably concise chemical names for the various substances with vitamin K activity. All the vitamins K are derivatives of menadione (2-methyl-1,4-napthoquinone) [7.24] or menadione itself. Phylloquinone [7.25], referred to as vitamin K_1 has a phytyl side chain. The vitamin K_2 series, the menaquinones, have side chains of varying lengths, up to 13 isoprenoid units, but those having between four and 10 are most commonly encountered, particularly the compound referred to as 'vitamin K_2' which has seven [7.26].

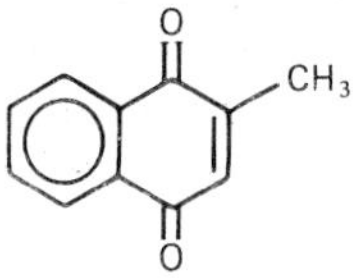

O CH$_3$ O

7.24

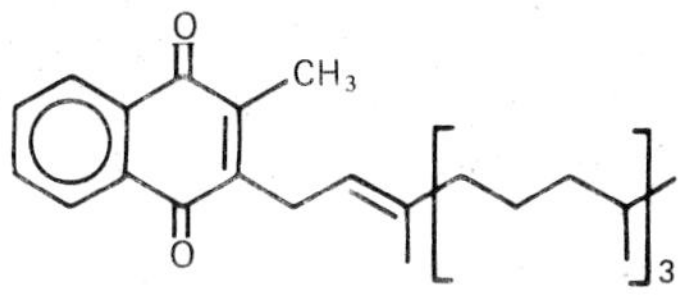

O CH$_3$ O 3

7.25

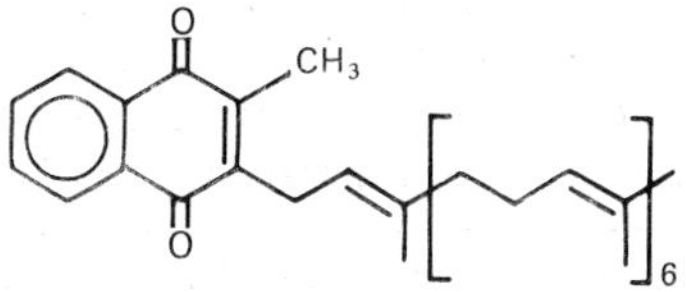

O CH$_3$ O 6

7.26

Vitamins K are widely distributed in biological materials in small amounts. Only leafy green vegetables such as spinach or cabbage are particularly rich in the vitamin (3–4 mg per 100 g). Other vegetables such as peas or tomatoes contain only 0.1–0.4 mg per 100 g. Animal tissues, including liver, contain similar levels. Cow's milk and human milk contain 2 and 20 μg per 100 g respectively. Although animals are unable to synthesise this vitamin humans are not particularly dependent on dietary supplies. This is because considerable quantities are synthesised by the bacteria of the large intestine. Vitamin K_2 was first isolated from putrefying fish meal. Deficiency of vitamin K is an important problem in poultry but deficiency is not encountered in otherwise healthy adult humans. Very occasionally new born infants do show deficiency symptoms.

The only physiological role for vitamin K that has been clearly identified is in blood clotting. The process of clot formation (the details are outside the scope of this book) depends upon the conversion of a number of different protein factors from inactive to active forms. The best understood of these is the conversion of the inactive *prothrombin* to the active proteolytic enzyme *thrombin*. This conversion requires the carboxylation of a pair of glutamate residues at one end of the polypeptide chain. A vitamin K molecule in its reduced form, *ie* a hydroquinone:

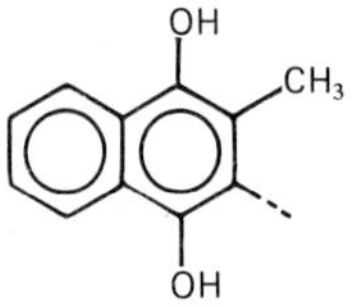

OH
CH_3
OH

is an essential component of the carboxylation reaction system. Vitamin K deficiency therefore results in a failure of the blood clotting mechanism. Many substances which antagonize the activity of vitamin K are now known. Most are derivatives of coumarin such as dicoumarol [7.27] (3,3′-methylene-*bis*-4-hydroxycoumarin) and warfarin [7.28] (3-(α-acetonylbenzyl)-4-hydroxycoumarin). Dicoumarol is produced when clover for animal feeding is spoiled by fermentation and causes disease in cattle. Warfarin is used as a rat poison.

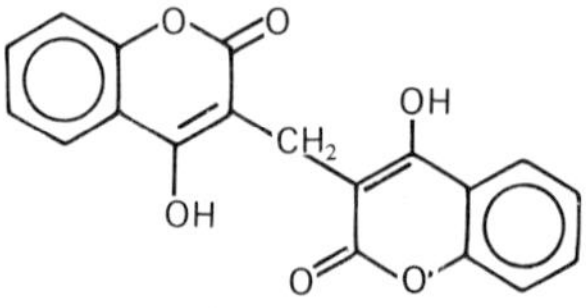

O O OH CH_2 OH O O

7.27

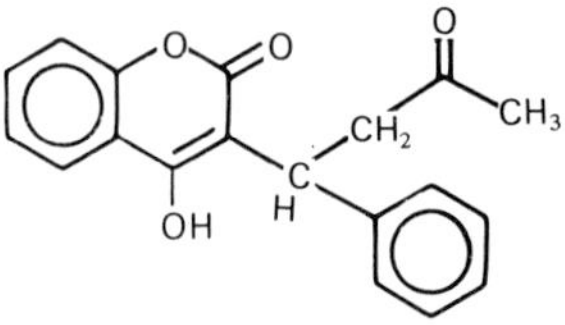

O O O CH_2 CH_3 C H OH

7.28

Vitamin K analysis is usually accomplished by thin layer chromatography of lipid extracts. The vitamin K activity of animal feeding stuffs has attracted far more attention than that of human foods. It appears that the vitamin is fairly stable under the usual conditions of food processing.

Further reading

A. E. Bender, *Food processing and nutrition*. London: Academic Press, 1978.

Nutrition reviews' 'Present knowledge in nutrition' (D. M. Hegsted *et al* eds), 4th Edition. New York: The Nutrition Foundation, 1976.

'Water soluble vitamins, hormones and antibiotics', Vol. 11 in *Comprehensive biochemistry* (M. Florkin and E. M. Stotz eds). London: Elsevier, 1963.

J. Marks, *The vitamins in health and disease*, London: Churchill, 1975.

'The fat soluble vitamins', Vol. 2 in *Handbook of lipid research* (H. F. DeLuca ed). New York: Plenum, 1978.

R. W. McGilvery, *Biochemistry, a functional approach*. Philadelphia: Saunders, 1979.

R. J. Taylor, *Micronutrients*. Unilever Educational Booklet No. 9, Unilever Limited, London, 1972.

Vitamin C, ascorbic acid (J. N. Counsell and D. H. Horning eds). London: Applied Science Publishers, 1981.

8. Preservatives

Micro-organisms, particularly bacteria, yeasts and moulds, have nutritional requirements remarkably similar to our own. Unless they use photosynthesis (as some bacteria do) energy is obtained by the oxidation or fermentation of organic compounds. Similarly organic compounds are the source of carbon and nitrogen for the biosynthesis of cell material. It is hardly surprising therefore that few types of micro-organisms will decline to utilize the same nutrients that make up a typical human diet – given the opportunity. The popularity of human foods with micro-organisms is enhanced by their tendency to be at moderate pH values, mild temperatures and, in one sort of food or another, to offer a wide enough range of oxygen tensions to tempt obligate anaerobes and obligate aerobes as well as less fastidious types.

Thus it is to be expected that a major goal of food technology has always been the *control* of food-borne micro-organisms. The word 'control' is stressed simply because the elimination of micro-organisms from food is frequently not even attempted and in some foodstuffs will be totally undesirable. In fact there are four quite distinct aspects of food science to which microbiologists contribute. The most important is that of safety. Foodstuffs are inevitably ideal carriers of pathogenic bacteria, *eg Salmonella* sp. as well as being ideal substrates for the growth of bacteria and fungi which secrete toxins, *eg Clostridium botulinum* and *Staphylococcus aureus*.

A second aspect of food microbiology is that of biodeterioration. If food materials were not vulnerable to degradation by the extracellular enzymes secreted by invading bacteria and moulds there would be little prospect of success for the parallel processes of digestion that occur in our own alimentary canals.

The second two aspects of food microbiology are less negative in outlook, *ie* where microbial activity is utilized in food production. Many processes, often under the blanket term 'fermentation', have been practised for thousands of years. For example cheese making and pickling are both exploitations of 'natural' processes of biodeterioration. Microbial action reduces the raw material to a more stable product where there is much less scope left for further breakdown. This is usually because microbial activity has been accompanied by a decrease in water activity and/or the accumulation of ethanol, lactic acid or acetic acid (ethanoic acid).

The fourth aspect of food microbiology, the most recent to attract interest, is the use of the micro-organisms themselves as food. While we have always eaten fungi such as mushrooms, great interest is now being shown in the use of single celled micro-organisms. These can be utilized directly but most often the protein is extracted and used as a food supplement for livestock. Strains of micro-organisms have been developed which grow readily on methanol and other by-products of the oil industry. The recent rises in world oil prices have unfortunately cast doubts on the economics of SCP, 'Single cell protein', production.

Of these various aspects of food microbiology the use of chemical substances to deter unwanted microbial activity is of greatest interest to chemists. Even with regard to unwanted micro-organisms the inhibition of growth is often the most one may hope to attain. As we have seen elsewhere in this book the heat processing methods required to obtain sterility, *ie* the absence of viable vegetative cells or spores, cannot be regarded as compatible with maximum nutritive value. Furthermore the essential similarity between the cellular components and metabolic processes of micro-organisms and those of man will ensure that chemical sterilants (*eg* chlorine or phenol in high concentrations) will hardly be acceptable as food ingredients. The advantage of a chemical additive approach to the reduction of microbial activity is that many products will need to remain stable for some time after the package has been opened. A jar of jam or a bottle of tomato ketchup will be subjected to repeated and massive recontamination once they come under the influence of the junior members of a household.

It is actually quite unusual for a single antimicrobial procedure to be used alone to protect a food product. Wherever possible several relatively mild procedures are combined to maximise the inhibition of microbial activity while minimising adverse effects on nutritional value of acceptability. For example the long term storage of meat, fish and vegetables at ambient temperatures demands the use of canning with the temperature, at the centre of the can, being held above 115 °C for over one hour. Otherwise the surviving spores of *Cl. botulinum* would find the anaerobic, nutrient rich, neutral pH environment ideal for germination, growth and toxin production. However, fruit's natural acidity eliminates the risk of botulism and the more modest heat treatments we associate with home bottling are adequate. As we shall see in the accounts of various food preservatives that follow they are very often combined in use with other antimicrobial procedures. For example the safety of cooked ham depends on (*i*) salt content to maintain a low water activity, (*ii*) cooking to destroy most vegetative bacterial cells and some spores, and (*iii*) nitrite to prevent spore germination and bacterial growth.

The first antimicrobial chemical to be used was undoubtedly salt, sodium chloride. One can be confident that it was its use as a preservative, rather than as a flavouring, that gave it its value to early civilisations. Salting is the traditional method of preserving meat, often in combination with smoking and drying. Modern technology may have provided more rapid methods of getting the salt into the meat but the essentials have remained unchanged for centuries. Salt solutions containing 15–25 per cent salt are used to bring the water activity, a_w, down to about 0.96.* This has the effect of retarding the growth of most meat spoilage organisms. With the advent of other preservative methods, notably canning and refrigeration, the importance of salting has diminished, but not without some salt-preserved meat products such as ham and bacon becoming firmly established in our diet. As the necessity for high salt content for preservation has lessened, public taste has adapted to less salty and less dry bacon and ham.

Nowadays very little meat is preserved by the use of common salt alone. At some unknown point in history it was realised that it was the unintended presence of saltpetre (sodium nitrate) as an impurity in the crude salt, that resulted in an attractive red or pink colour in preserved meat. Subsequently nitrates and/or nitrites have become an almost indispensible component of the salt mixtures (known as 'pickles') used for curing bacon and ham. In spite of the antiquity of the process it is only very recently that the special antimicrobial properties of the nitrite in ham have been recognised. There are many subtle variations in curing procedures but the well known 'Wiltshire cure' is typical. Sides of pork are injected with about 5 per cent of their own weight of a pickling brine containing 25–30 per cent sodium chloride, 2.5–4 per cent sodium or potassium nitrate and sometimes a little sugar. The sides are then submerged in a similar solution for a few days. After removal they are stored for 1–2 weeks for 'maturation' to occur. If prolonged storage is then called for a second preservative, smoke, is applied.

By the time the maturation stage is completed the salts will be evenly distributed throughout the muscle tissue and a complex series of reactions will have given rise to the characteristic red colour of *uncooked* bacon or ham. The essential features of these reactions can be summarized as follows:

---

*The use of sugar to achieve a similar result in fruit preservation was discussed in Chapter 2.

(*i*) Some of the nitrate present is reduced to nitrite:

$$NO_3^- + 2[H] \rightarrow NO_2^- + H_2O$$

either by salt-tolerant micro-organisms in the brine or by the respiratory enzymes of the muscle tissue.
(*ii*) The nitrite oxidises the iron of the muscle myoglobin to the iron(III) state (see Chapter 4):

$$Fe(II) + NO_2^- + H^+ \rightarrow Fe(III) + NO + OH^-$$

ie myoglobin (Mb) is converted to metmyoglobin (MMb).
(*iii*) The resulting nitrogen oxide reacts with the iron of MMb to form nitrosyl metmyoglobin (MMbNO).
(*iv*) The MMbNO is immediately reduced by the respiratory systems of the muscle tissue to nitrosyl myoglobin, MbNO, the red pigment of uncooked bacon and ham.

The distribution of electrons around the iron of MbNO is similar to that in oxymyoglobin (MbO_2), hence the similarity in colour. When bacon is grilled or fried and when ham is boiled, the nitrosyl myoglobin is denatured and a bright pink pigment, often referred to as nitrosylhaemochromogen, is formed. There is no certainty as to its structure but it is believed that the denaturation of the globin allows a second nitrogen oxide molecule to bind to the iron in the place of the histidine residue.

In recent years the use of nitrates in food has been regarded with increasing suspicion due to the risk of nitrosamine formation. Nitrosamines are potent carcinogens. Secondary amines react readily with nitrous acid to form stable *N*-nitroso compounds:

$$R'R''NH + NO_2^- \longrightarrow R'R''N{-}N{=}O + OH^-$$

The reaction with primary amines, which are of course abundant in meat as free amino acids, leads simply to deamination:

$$R{-}NH_2 + NO_2^- \rightarrow N_2 + ROH + OH^-$$

The reaction with amino acids is important in that it will lead to some elimination of the excess nitrite in cured meat. Secondary amines are much less abundant in meat but are assumed to arise as a result of microbial action, especially by anaerobes. The decarboxylation of proline, which occurs spontaneously at the high temperatures involved in frying, leads to the formation of the very important nitrosamine, *N*-nitrosopyrrolidine (NOPyr):

$$\underset{\text{(ring: }H_2C{-}CH_2,\ H_2C,\ CH{-}COOH,\ NH)}{\text{proline}} \xrightarrow{-CO_2} \underset{\text{(ring: }H_2C{-}CH_2,\ H_2C,\ CH_2,\ NH)}{\text{pyrrolidine}} \xrightarrow{NO_2^-} \underset{\text{(ring: }H_2C{-}CH_2,\ H_2C,\ CH,\ N{-}NO)}{\text{NOPyr}}$$

Although nitrosamines have been detected in many cured meat products such as salami and frankfurters it is clear that they are most significant in cured meat that has been cooked at high temperatures, such as fried bacon. Levels of NOPyr in fried bacon are consistently found to be around 100 μg kg^{-1} with rather lower levels of *N*-nitrosodimethylamine. Although there is no doubt as to the carcinogenicity of these volatile nitrosamines when tested in laboratory animals, there is no evidence that the consumption of cured meats has actually been responsible for disease in man. Nevertheless the potential hazard is sufficient to ensure that steps are being taken to reduce the incidence of *N*-nitrosamines in all cured meat products.

Improvements in processing controls are making possible a steady reduction in residual nitrite levels but the most important measure is the inclusion of ascorbic acid in curing salt mixtures. Ascorbic acid is beneficial in two ways. Firstly, being a reducing agent, it enhances, directly or indirectly, the rates of the key reducing reactions in the formation of MbNO, thereby allowing lower levels of nitrites or nitrates to be used in the pickles. Secondly it actually inhibits the nitrosation reaction.

Even the remote possibility of a hazard in the use of nitrite for curing would justify its prohibition if its sole value was a colouring agent. (The question of whether nitrite actually contributes to the flavour of cured meat remains unresolved.) However, its antimicrobial properties would justify its inclusion even if it had no effect on colour. It has been known for some years that the growth of many types of anaerobic bacteria, including the causative organism of botulism, *Cl. botulinum*, is prevented in cured meat by an unidentified product of the interaction of the residual nitrite with the meat, that arises when the meat is cooked. If ham was given a sufficient heat treatment to ensure that all spores of *Cl. botulinum* had been killed it would be unacceptably overcooked. Canned stewing beef, which is given such a heat treatment, provides a good illustration of the sort of texture the ham would have. This effect of the residual nitrite in ham was termed the 'Perigo effect' after its discoverer and a great deal of work was done to identify the 'Perigo inhibitor'. It is now well established that during cooking much residual nitrite is broken down to nitrogen oxide. This is not liberated from the meat but becomes weakly associated with some of the exposed amino acid side chains and iron atoms of the denatured meat proteins. From this reservoir nitrogen oxide is available for the inibition of sensitive bacteria. It appears that nitrogen oxide is a specific, and potent, inhibitor of an enzyme whose activity is crucial to the energy metabolism, and therefore growth and toxin production, of anaerobes such as *Cl. botulinum*.

The use of nitrite therefore presents those concerned with food safety with a paradox. On the one hand we have nitrite the preservative which has been used for centuries and is now known to be essential for microbiological safety, and on the other hand we have nitrite the colouring agent suspected of giving rise to carcinogens. Apart from the obvious but unacceptable* solution of eliminating cured meats from our diet the most we can do is to ensure that residual nitrite levels are never much more than the minimum, about 50 μg per gram, needed to prevent toxin production. It is also worth remembering that nitrate naturally present in other foods, particularly vegetables, and in drinking water, can be reduced to nitrite by our intestinal bacteria and thereby cause problems even for those who never eat cured meat.

Smoke is the other preservative traditionally associated with meat and fish. We can be fairly sure that the flavouring action of wood smoke was initially regarded only as a valuable side effect of drying out over a wood fire. The preservative action of smoke most certainly went unnoticed. Nowadays meat and fish are rarely preserved by drying, and refrigeration has made the preservative action of smoke less important than its flavour.

Smoke consists of two phases, a disperse phase of liquid droplets and a continuous gas phase. In smoking, the absorption of gas phase components by the food surface is considered to be much more important than the actual deposition of smoke droplets. The gas phase of wood smoke has been shown to include over 200 different compounds including formaldehyde (methanal), formic acid (methanoic acid), short chain fatty acids, vanillic and syringic acids [8.1 and 8.2], furfural [8.3], methanol, ethanol, acetaldehyde (ethanal), diacetyl (butanedione), acetone (propanone) and 3,4-benzpyrene [8.4]. Of these the most important antimicrobial compound is almost certainly formaldehyde (methanal).

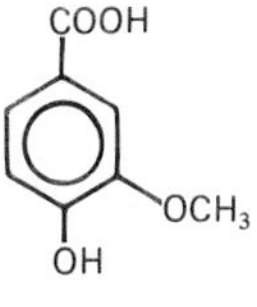

8.1

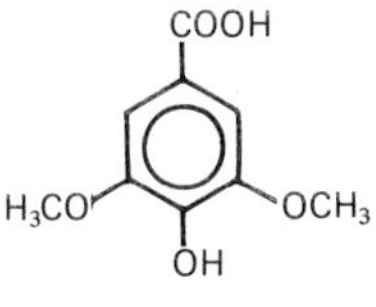

8.2

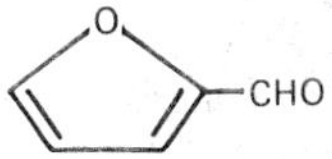

8.3

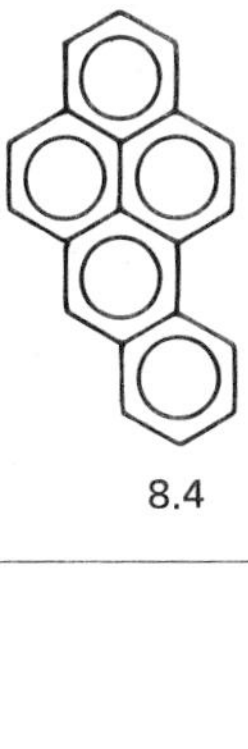

8.4

* At least to this author.

The detection of known carcinogens such as 3,4-benzpyrene and other polynuclear aromatic compounds in wood smoke has led to concern over the safety of smoked foods. It has been suggested that the high incidence of stomach cancer in Iceland and the Scandinavian countries is due to the large amounts of smoked fish consumed there. Although there is no suggestion that the amount of smoked food consumed elsewhere in the world is sufficient to cause similar problems liquid smoke preparations which do not contain the polynuclear aromatics are being adopted to an increasing extent. Liquid smokes are prepared from wood smoke condensates by fractional distillation and water extraction. The undesirable polynuclear hydrocarbons are not soluble in water.

It is sad that so many consumers overlook the fact that these modern developments in curing and smoking have been introduced to reduce the small, but very real, risks involved in traditional processes and that with both smoke and nitrite it is the preservative effect rather than flavour or colour that commends them to the food processor.

Sulphur dioxide (SO_2) has been used in wine making for hundreds of years to control the growth of unwanted micro-organisms. It was originally obtained by the rather haphazard process of burning sulphur and exposing the *must*, *ie* the unfermented grape juice, to the fumes. Nowadays the free gas is rarely used, being replaced by a number of SO_2 generating compounds, particularly sodium sulphite (Na_2SO_3), sodium hydrogensulphite ($NaHSO_3$) and sodium metabisulphite (sodium disulphite) ($Na_2S_2O_5$). The relationships between these, sulphur dioxide and sulphurous acid are outlined in *Fig. 8.1*. The antimicrobial activity

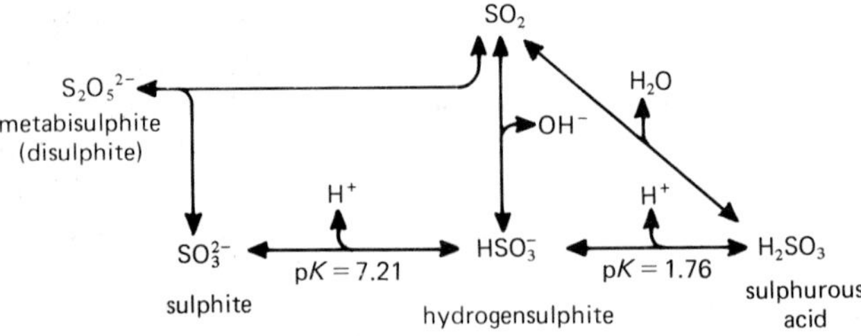

FIG. 8.1. Structural relationships of sulphur dioxide-generating compounds.

of these compounds increases dramatically as the pH falls and it is therefore assumed that it is the undissociated sulphurous acid that has the antimicrobial activity. Total sulphur dioxide levels around 100 ppm are added to musts to achieve a differential effect – the desirable wine yeast *Saccharomyces cerevisiae* is able to grow and ferment the sucrose to ethanol but some undesirable yeast species such as *Kloeckera apiculata* as well as lactic acid producing bacteria are suppressed.

About two-thirds of the total sulphur dioxide in a must or wine is bound to anthocyanins and other flavanoids (see Chapter 5), sugars:

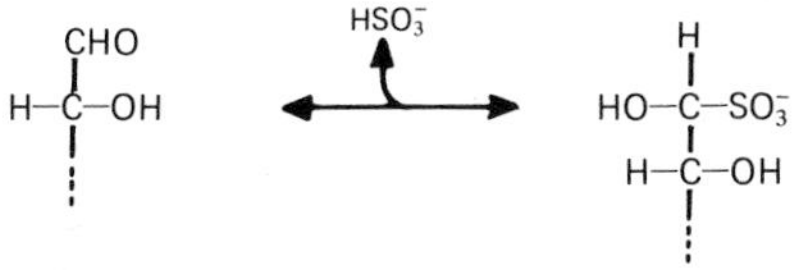

$$\begin{array}{c} CHO \\ | \\ H-C-OH \\ | \\ \vdots \end{array} \xrightarrow{HSO_3^-} \begin{array}{c} H \\ | \\ HO-C-SO_3^- \\ | \\ H-C-OH \\ | \\ \vdots \end{array}$$

and other aldehydes. Besides the sulphur dioxide that is added many strains of wine yeasts actually produce sulphur dioxide themselves by the reduction of sulphate present in the grape juice. When the wine is bottled further sulphur dioxide is added to prevent secondary fermentation, *ie* fermentation of residual sugar in the bottle; concentrations of total sulphur dioxide up to around 500 ppm are commonly used.

A wide range of other foodstuffs, particularly fruit or vegetable based products, have sulphite added as a preservative or as a residue from processing operations. The use of sulphite as an antioxidant has already been examined (Chapter 5).

The use of sulphur dioxide is generally regarded as being without any toxicity hazard at the usual levels but it does pose an important nutritional problem. The bisulphite (hydrogen sulphite) ion reacts readily, and destructively, with the vitamin thiamine (see Chapter 7). Vegetables such as potatoes which are often stored in sulphite solutions at intermediate stages of processing will lose considerable proportions of their thiamine content. It is for this reason that many countries prohibit the use of sulphur dioxide in foodstuffs that are important sources of thiamine in the diet.

A further drawback to the use of sulphite/sulphur dioxide is the taste. Above 500 ppm most people are aware of its disagreeable flavour and some can detect it at much lower levels. Some white wines are actually characterised by their slight sulphur dioxide flavour.

Very little is known for certain about how sulphur dioxide inhibits microbial growth. The most probable explanation is that addition compounds are formed with the aldehyde groups of key metabolic intermediates or coenzymes, but these have not been identified.

Benzoic acid [8.5] occurs naturally in small amounts in some plant foods but it is the synthetic product that is widely used as a food preservative. The more soluble sodium salt is the form in which it is usually added to food but it is the undissociated acid that is active against micro-organisms, particularly yeasts and bacteria. Since it is the protonated form that is active benzoic acid use is restricted to acid

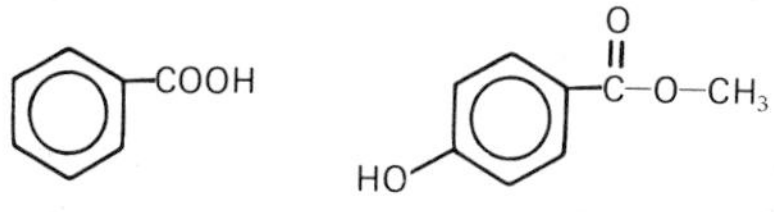

C_6H_5-COOH

8.5

$HO-C_6H_4-C(=O)-O-CH_3$

8.6

foods (pH 2.5–4.0) such as fruit juices and other beverages. At the levels used, 0.05–0.1 per cent, no deleterious effects on humans have been detected. Benzoate is not accumulated in the body but is converted, by condensation with glycine, to hippuric acid (*N*-benzoylglycine) which is excreted in the urine:

$$C_6H_5-COOH + H_2N-CH_2-COOH \xrightarrow{\uparrow H_2O} C_6H_5-\overset{O}{\overset{\|}{C}}-\underset{H}{N}-CH_2-COOH$$

Esters of *p*-hydroxybenzoic acid with methanol [8.6], propanol and other alcohols, known collectively as 'parabens' are also commonly used in most of the same situations as benzoic acid and present similarly little problem of toxicity.

Another organic acid of increasing popularity as a preservative is sorbic acid [8.7]. Sorbic acid, included as such or as its sodium or potassium salt, is a particularly effective inhibitor of mould and yeast growth. As with benzoic acid it is the undissociated acid that is the active form but the slightly higher pK_a value (4.8 compared with 4.2) means that it can be effective at higher pH values, even up to pH 6.5. It can therefore be used in a much wider range of food products including processed cheese and flour confectionery (*ie* cakes and pastries but not bread). At the levels normally used (up to 0.3 per cent) no toxic effects have been detected. It is suggested that it is readily metabolized by mammals by the same route as naturally occurring unsaturated fatty acids.

$$CH_3-CH{=}CH-CH{=}CH-COOH$$

8.7

There is a surprising lack of information as to the mechanisms by which the organic acid preservatives actually inhibit microbial growth. The most probable site of action is the cell membrane where it may be imagined that these molecules would be able to form close associations with the polar membrane lipids. In doing so it is likely that they would disrupt the normal processes of active transport into the cell. Whether such a mechanism could apply to the preservative action of propionic acid, and its salts, which are permitted for use in baked goods, including bread, remains to be seen.

Further reading

Micro-organisms, form, function and environment (L. E. Hawker and A. H. Lindon eds). London: Edward Arnold, 1979.

Developments in food preservatives – 1 (R. H. Tilbury, ed). London: Applied Science, 1980.

R. J. Taylor, *Food additives*. Chichester: Wiley, 1980.

J. Stephen and R. A. Pietrowski, *Bacterial toxins*. Walton-on-Thames: Nelson, 1981.

R. A. Lawrie, *Meat science*, 3rd edn. Oxford: Pergamon, 1980.

Prescott & Dunn's industrial microbiology (G. Reed ed), 4th edn. Westport: AVI, 1982.

W. C. Frazier and D. C. Westhoff, *Food microbiology*, 3rd edn. New York: McGraw Hill, 1978.

Appendix 1. General food chemistry texts for further reading

Specialised texts have been listed at the end of each chapter but the following are relevant to a wider range of topics.

J. Hawthorne, *Foundations of food science*. Oxford: Freeman, 1981.

Z. Berk, *Braverman's introduction to the biochemistry of foods*. London: Elsevier, 1976.

N. A. M. Eskin, H. M. Henderson and R. J. Townsend, *Biochemistry of foods*. London: Academic, 1971.

W. Heimann, *Fundamentals of food chemistry*. Chichester: Ellis Horwood, 1980.

Principles of food science. Part I, Food chemistry (O. R. Fennema ed). New York: Dekker, 1976.

Effects of heating on foodstuffs (R. J. Priestley ed). London: Applied Science, 1979.

Nutritional and safety aspects of food processing (S. R. Tannenbaum ed). New York: Dekker, 1979.

H. Egan, R. S. Kirk and R. Sawyer, *Pearson's chemical analysis of foods*, 8th edn. London: Churchill Livingstone, 1981. (This is the authoritative source for laboratory techniques in relation to the food standards required by UK legislation.)

H. Charley, *Food science*, 2nd edn. New York: Wiley, 1982. (By far the most effective text at linking the findings of the laboratory to the work in the kitchen.)

Appendix 2. EEC numbers for food additives

Additives that are permitted for use in the EEC are given reference numbers that are frequently used on food labelling. This abbreviated version of the list of EEC numbers should aid the identification of many food product ingredients.

100	Curcumin
101	Riboflavin
102	Tartrazine
104	Quinoline Yellow
110	Sunset Yellow FCF
120	Cochineal
122	Carmoisine
123	Amaranth
124	Ponceau 4R
127	Erythrosine
131	Patent Blue V
132	Indigo Carmine
140	Chlorophyll
141	Chlorophyll derivatives
142	Food Green S
150	Caramel
151	Black PN
153	Carbon Black
160	Carotenes
161	Xanthophylls
162	Betalaines
163	Anthocyanins
170	Calcium carbonate
171	Titanium dioxide
172	Iron oxides and hydroxides
173	Aluminium
174	Silver
175	Gold
200	Sorbic acid
201–3	Sorbates (Na^+, K^+, Ca^{++})
210	Benzoic acid
211–3	Benzoates (Na^+, K^+, Ca^{++})
214–9	Esters of 4-hydroxy benzoate
220	Sulphur-dioxide
221–7	Sulphites and metabisulphites
249, 50	Nitrites (K^+, Na^+)
251, 2	Nitrates (Na^+, K^+)
260	Acetic acid (ethanoic acid)
261–3	Acetates (Na^+, K^+, Ca^{++})
270	Lactic acid
280	Propionic acid
281–3	Propionates (Na^+, Ca^{2+}, K^+)
290	Carbon dioxide
300	L-Ascorbic acid
301, 2	Ascorbates (Na^+, Ca^{++})
304	Ascorbyl palmitate
306	Natural tocopherols
307–9	Synthetic tocopherols
310–12	Gallate esters
320	Butylated hydroxyanisole
321	Butylated hydroxytoluene
322	Lecithins
325–7	Lactates (Na^+, K^+, Ca^{++})
330	Citric acid
331–3	Citrates (Na^+, K^+, Ca^{++})
334	Tartaric acid
335–7	Tartrates (Na^+, K^+, Na^+K^+)
338	Orthophosphoric acid
339–41	Orthophosphates (Na^+, K^+, Ca^{++})
400	Alginic acid
401–4	Alginates (Na^+, K^+, NH_4^+, Ca^{2+})

406 Agar
407 Carageenan
410 Locust bean gum
412 Guar gum
413 Gum tragacanth
414 Gum arabic (acacia)
420 Sorbitol
421 Mannitol
422 Glycerol
440 Pectin
442 Ammonium phosphatides
450 Oligo- and polyphosphates
460 Cellulose
461-6 Carboxymethyl cellulose and similar cellulose derivatives
470 Fatty acid salts
471 Mono- and diglycerides
472 Mono- and diglyceride esters with organic acids, *eg* citric
473 Sucrose esters of fatty acids
474 Sucroglycerides
481-2 Stearoyl-2-lactylates (Na^+, Ca^{++})
491-4 Sorbitan fatty acid esters

Index

Page numbers in italics refer to structural formulae

* and see particular named examples